Fundamentals of Molecular Bioengineering

Luisa Di Paola

Fundamentals of Molecular Bioengineering

Luisa Di Paola
Faculty of Engineering
Bio-Medico di Roma
Roma, Italy

ISBN 978-3-031-42024-5 ISBN 978-3-031-42022-1 (eBook)
https://doi.org/10.1007/978-3-031-42022-1

The original submitted manuscript has been translated into English. The translation was done using artificial intelligence. A subsequent revision was performed by the author(s) to further refine the work and to ensure that the translation is appropriate concerning content and scientific correctness. It may, however, read stylistically different from a conventional translation.

This Springer imprint is published by the registered company Springer Nature Switzerland AG
The registered company address is: Gewerbestrasse 11, 6330 Cham, Switzerland

Preface

Così tra questa immensità s'annega il pensier mio: e il naufragar mi è dolce in questo mare..

So in this immensity my thoughts drown: and sinking is so sweet to me in this sea.

G. Leopardi

The comprehension of life is one of the most formidable endeavors in the history of scientific inquiry. When we examine the intricacies and constituent elements of life, it can seem complex and convoluted. However, when we examine its behavior, life reveals a remarkable simplicity.

For example, we may not be able to exhaustively describe every molecular mechanism at work within a rabbit, but it is evident that the rabbit will instinctively flee when startled by a sudden noise (cit. Alessandro Giuliani). This captivating simplicity of behavior amidst the vast complexity of biological machinery captivates poets, scientists, and even engineers, who endeavor to replicate the actions of biological systems akin to "machines."

Adopting the mindset of inquisitive children, physicists and engineers embark on a path of dissecting objects to unravel their inner workings and behavioral patterns. However, this approach falls short when attempting, for instance, to dismantle a flower, as it yields only fragmented and irreparable components that can never be reassembled into a flower again.

While comprehending the individual components of life is crucial, it alone is insufficient to unravel the enigmatic entirety of life's mystery. The ideal world, which is described only in terms of the properties of components, excludes the possibility of interactions between the parts themselves, which act as a kind of additional element, other than the "parts" themselves, in real systems.

This systemic approach to the description of "real systems," when opposed to "ideal," is inherent in the chemical thermodynamics, when describing real systems in terms of properties (such as fugacity and activity fugacity) accounting for interactions between molecular components.

Adopting this mindset, this book attempts to provide a comprehensive perspective on "biomolecular engineering," including in this definition the molecular biophysics of biological systems and the most advanced techniques in computational biology.

Biomolecular engineering is a rapidly growing field that has the potential to revolutionize our understanding of life and to develop new and innovative technologies. This book provides a foundation for understanding the principles of biomolecular engineering and for exploring the potential of this field to address some of the most pressing challenges facing our world.

This book is just a small glimpse into the vast and fascinating world of biomolecular engineering. I hope that it will inspire you to learn more about this field and to contribute to its future development.

As the verses at the beginning of this preface say, this is an immensity. I've just offered a small glance behind a small hole in the fence. Yet, even this small glimpse reveals a sight that is still so fascinating to me.

Roma, Italy Luisa Di Paola
July 13, 2023

Contents

Part I Biomolecular Thermodynamics

1 Classical and Chemical Thermodynamics 3
1.1 Basic Definitions 3
1.2 Principles of Thermodynamics 4
1.2.1 First Principle of Thermodynamics and Internal Energy 5
1.2.2 Second Principle of Thermodynamics and Entropy 7
1.3 Thermodynamic Potentials and Chemical Equilibrium 9
1.3.1 Legendre Transformation and Thermodynamic Potentials 9
1.3.2 Partial Molar Properties and Chemical Potential 11
1.3.3 Fugacity, Chemical Potential, and Activity 14
1.3.4 Thermodynamic Phase Equilibria 16

2 Statistical Thermodynamics 19
2.1 Introductive Elements 19
2.2 Kinetic Theory of Gases 20
2.2.1 Molecular Interpretation of a Gas Pressure and Relationship with Internal Energy 21
2.2.2 Relationship Between Temperature and Molecular Properties 22
2.3 Principles of Statistical Mechanics 23
2.3.1 Distribution of Molecular Velocities 25
2.3.2 Application of Kinetic Theory of Gases 27
2.4 Ensembles and Postulates in Statistical Thermodynamics 31
2.4.1 Canonical Ensemble 34
2.4.2 Grand Canonical Ensemble 40
2.4.3 Microcanonical Ensemble 41
2.4.4 Microscopical Interpretation of Entropy 43

3 Proteins 47
3.1 Introductive Elements 48
3.2 The Amino Acids 48

3.3 Protein Structure 49
3.3.1 Primary Structure 51
3.3.2 Secondary Structure 51
3.3.3 Tertiary and Quaternary Structure 52
3.4 Protein Function and Relationship with Structure 54
3.5 Protein Folding and Denaturation 57
3.5.1 Denaturing Agents 62
3.5.2 Cooperativity and Folding 64

4 Intermolecular Forces and Potentials 67
4.1 Introduction 67
4.2 Definition of Intermolecular Potentials 70
4.3 Structure and Properties of Liquids 73

5 Volumetric Properties of Fluids: Equations of State 75
5.1 Introduction 75
5.2 Equations of State for Ideal and Nonideal Systems 76
5.3 Osmotic Pressure of Colloidal Systems 78
5.4 Virial Coefficients and Intermolecular Potentials: MacMillan-Mayer Theory 79
5.4.1 Virial Equation of State for Osmotic Pressure 83
5.5 Virial Coefficients of the Osmotic Pressure in Classical Thermodynamics 84

6 Biopolymer Ionization and Salting-Out 91
6.1 Introduction 91
6.2 Ionization Equilibria for Amino Acids, Proteins, and Peptides 92
6.2.1 Ionization Equilibria of Residues in Peptides and Proteins 96
6.3 Salting-Out and Phase Equilibria of Solutions of Biomacromolecules 98
6.3.1 Protein Salting-Out: The Hofmeister Series 100
6.3.2 An Equation of State for Colloidal Solutions 101
6.3.3 Phase Equilibria for Biomacromolecules 103
6.3.4 Molecular Thermodynamics of Solvation of Biomacromolecules 104

7 Protein-Ligand Interactions and Binding 107
7.1 Introductive Elements 107
7.1.1 Simple Interaction: Single Class of Independent Binding Sites 109
7.1.2 Multiple Sites, Allosteric Interactions, and Competitive Binding 119
7.2 Molecular Models for Allostery 122
7.2.1 The Allosteric Regulation of the Blood Hemoglobin 126

7.3 Molecular Models of Enzymatic Reactions 128
7.3.1 Mechanism of Enzymatic Activity 130
7.3.2 Enzyme Inhibition 134
7.3.3 Noncompetitive Inhibition 136

Part II Computational Biology

8 Molecular Dynamics Simulations for Computational Biology 141
8.1 Introduction 141
8.2 Basic Concepts in Molecular Dynamics 142
8.2.1 Design Constraints and Ensembles 143
8.2.2 Molecular Simulations Force Fields and Potentials 153
8.2.3 Coarse-Grained Molecular Dynamics of Proteins 155
8.3 Computational Tools for Molecular Dynamics 159

9 Protein Structure Prediction and Analysis 163
9.1 The Computational Approach to the Protein Folding Problem 163
9.1.1 The Theory of Protein Folding 164
9.1.2 Werewolves, Shapeshifters, and Replicants: Strange Figures Among the Protein Community 168
9.2 Computational Approaches and Tools for Protein Folding Prediction 171
9.3 Sequence Alignment and Homology Modeling 174
9.3.1 The Protein Sequence Alignment Problem and Its Applications 174
9.3.2 Computational Tools for Homology Modeling of Protein Structure 176
9.4 Computational Prediction of Protein Stability 180
9.4.1 Theoretical Description of Protein Stability 180
9.4.2 Computational Tools for Protein Stability Prediction 182
9.4.3 Protein Stability Prediction in Quaternary Structure 184

10 Systems Biology and Structure-Based Protein Networks 187
10.1 Introduction 187
10.2 Basic Definition of Theory of Complex Systems 188
10.2.1 Complex Systems and Networks 189
10.3 Systems Biology: Biological Entities as Complex Systems 196
10.4 Structure-Based Protein Networks 199
10.4.1 Protein Contact Networks 201
10.5 Dynamics-Based Protein Networks 212
10.5.1 Elastic Network Modeling 213
10.5.2 Energy Flow Networks 217

11 Computational Protein Binding 223
11.1 Introduction 223
11.2 Computational Prediction of Protein-Ligand Binding Sites 224

11.2.1 Experimental Methods and Theoretical Modeling of Protein Binding 226
11.2.2 Computational Approaches for Prediction of Protein-Ligand Binding Sites 229
11.3 Computational Approach to Allostery 236
11.3.1 Network-Based Computational Approaches for Allostery Quantitative Identification 238
11.4 Computational Analysis of Protein-Protein Interactions 249
11.4.1 Theoretical Analysis of Protein-Protein Interfaces 252
11.4.2 Computational Approach to the Prediction and the Analysis of Protein-Protein Interactions 257

Bibliography 265

Index 267

Part I

Biomolecular Thermodynamics

Classical and Chemical Thermodynamics 1

Nothing in life is certain except death, taxes and the second law of thermodynamics.

S. Lloyd

Abstract

Why it is important to know this material?

Chemical thermodynamics are the very foundation of thermodynamic phase equilibria, which play a central role in biochemical and biotechnology systems.

What is the key idea?

The thermodynamic equilibria for each component in multicomponent mixtures can be defined by starting from the definition of chemical potential and fugacity, passing by fugacity and activity.

What is necessary to know already? It is important to have a clear idea of the principles of thermodynamics and define thermodynamic potential (internal energy, entropy, enthalpy, free energy).

1.1 Basic Definitions

Classical thermodynamics is based on the definition of general principles and rules, for the analysis of transformations of regions of space; in particular, thermodynamics properly deals with the transformation between equilibrium states of such regions, called *systems*.

A thermodynamic system is a portion of space bounded by a boundary, real or virtual, that separates it from the (external) environment. Based on the exchanges that occur with the environment, the following characteristic systems can be defined:

L. Di Paola, *Fundamentals of Molecular Bioengineering*,
https://doi.org/10.1007/978-3-031-42022-1_1

- **Isolated** system: It exchanges neither matter nor energy with the environment.
- **Closed** system: It exchanges energy but not matter with the environment; in particular, systems that do not exchange heat with the outside world are called adiabatical systems.
- **Open** system: It exchanges both energy and matter with the environment.

The *universe* is the whole including both the system and the external environment. The term *state of a system* identifies the set of quantities, called state variables, that define the system under the assigned conditions; the variables are specifically chosen to represent the problem of interest. Directly measured quantities are called *primitives*; if they are evaluated from other quantities, they are called *derivatives*. The quantities can be grouped into two groups:

- **Intensive** quantities: Their value does not depend on the size of the system (e.g., temperature).
- **Extensive** quantities: They depend on the mass present in the system (e.g., volume).

Intensive quantities are also extensive quantities per unit volume or mass; in this case, they are called *specific* quantities.

Transformation of a system refers to the change in the value of state variables over time as a result of changing external conditions (e.g., temperature); if an inversion of the introduced perturbation allows an inversion of the function describing the evolution over time, the transformation is called *reversible* process, otherwise *irreversible*.

The term *phase* refers to a part of the system that is mechanically separable from the rest; when there are several phases, in a system at thermodynamic equilibrium, there is a relationship between the properties of these individual phases, which can be considered homogeneous throughout the reference volume of the individual phases; this relationship is dictated by the law of thermodynamic equilibrium between phases.

1.2 Principles of Thermodynamics

The laws of thermodynamic equilibrium, described for systems of various types, fall into the broader generality of the basic principles of classical thermodynamics, which establish constraints between the properties of systems that are in equilibrium. In particular, with reference to the first principle of thermodynamics, a general criterion of conservation of energy is applied to describe the transformation.

1.2.1 First Principle of Thermodynamics and Internal Energy

At first, it is necessary to identify the causes that generate the change in the state of a system, which give rise to the transformation between two equilibrium conditions: the change can be determined by an exchange of heat, mechanic energy, or by a change in the chemical composition of the system itself. Accordingly, the energy change ΔE_{AB} associated with the transformation $A \rightarrow B$ comprises three terms, each of which can be associated with the above contributions:

$$\Delta E_{AB} = \underbrace{\Delta E_{\mathscr{Q}}}_{\text{heat}} + \underbrace{\Delta E_{\mathscr{L}}}_{\text{work}} + \underbrace{\Delta E_{\mathscr{C}}}_{\text{chemical energy}} \tag{1.1}$$

The characteristic energy, which sums up all these contributions, is called the *internal energy* and corresponds to the sum of the characteristic energy for each individual molecule constituting the system; it is generally denoted as ΔU and is given as:

$$\Delta U = \delta\mathscr{L} + \delta\mathscr{Q} \tag{1.2}$$

where $\delta\mathscr{L}$ and $\delta\mathscr{Q}$ represent, respectively, the changes in mechanical work and heat associated with the assigned transformation $A \rightarrow B$

Internal energy U is a state function, that is, the change in U associated with an A → B transformation $\Delta U_{A\rightarrow B}$ is a function only of the state of the system at states A and B:

$$\Delta U_{A\rightarrow B} = U(B) - U(A) \tag{1.3}$$

For an infinite transformation of state variables, the associated internal energy change dU is an exact differential, since the function U (state variables) is known: this property is a consequence of the fact that U is a state function. On the other hand, as shown in Fig. 1.1, heat Q and work L do not have an unambiguous dependence from the state variables, so their variation depends on the type of transformation, and not only on the initial and final conditions that distinguish the transformation; this feature is represented through the notation δ, which denotes the variation of a function that does not correspond to an exact differential, denoted by d:

$$dU = \delta\mathscr{Q} + \delta\mathscr{L} \tag{1.4}$$

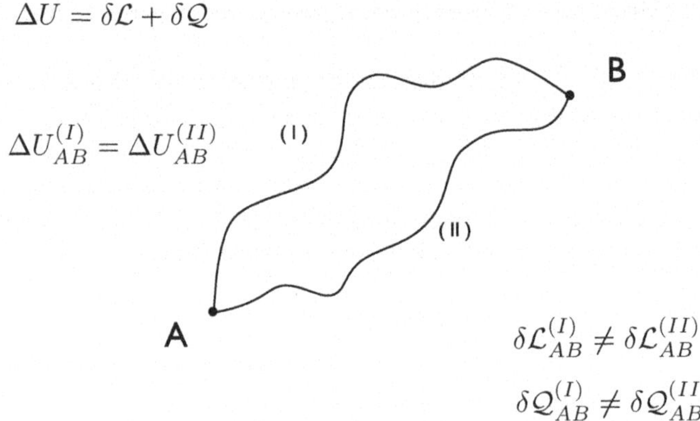

Fig. 1.1 First principle of thermodynamics: internal energy is a state function

In other words, $\mathcal{Q}$ and $\mathcal{L}$ are not functions of state, but their sum (internal energy U) is; a consequence of the first principle of thermodynamics is that the variation of internal energy for a cyclic process—that is, in which the initial and final points coincide—is zero:

$$\oint dU = 0 \tag{1.5}$$

This formulation, actually attributed to Max Planck, enshrines the impossibility of generating perpetual motion:

> In no way, neither through mechanical nor thermal, chemical, or other processes, is it possible to achieve perpetual motion... In other words, it is impossible to construct an engine that operates in a cycle and produces continuous work or kinetic energy from nothing.

Although the definition in the form of imperfect differentials for work and heat prevents a generalization of expression as a function of state variables, in any case the two terms associated with an infinitesimal transformation sometimes have well-defined expressions as a function of state variables. For example, in the case of discontinuous systems, mechanical work can be associated with volume changes that can be associated with changes in pressure and temperature; in particular, if P

is the internal pressure of the system, which remains constant in the transformation, mechanical work will be given by:[1]

$$\delta\mathscr{L} = -PdV \tag{1.6}$$

In this expression, the sign takes into account the so-called selfish convention, i.e., all energy contributions (mechanical work + heat) acquired by the system under consideration are considered positive; in this case, a positive change in dV implies an expansion and, therefore, a work exerted by the system on the outside, thus of negative sign, overall; in expression 1.6, the functional relationship between P, V, and T is provided by the equations of state, which will be discussed extensively later; e.g., the work produced for an isothermal transformation, at a fixed temperature T, by the n moles of a perfect gas ($V = nRT/P$):

$$\delta\mathscr{L} = P\frac{nRT}{P^2}dP \quad \Longrightarrow \mathscr{L} = nRT \cdot \log\left(\frac{P_2}{P_1}\right) \tag{1.7}$$

1.2.2 Second Principle of Thermodynamics and Entropy

The second principle of thermodynamics was formulated from the analysis of the experimental results of Sadi Carnot, who devoted himself to the practical study of the transformation of heat into work, through thermal cycles involving a system in contact with heat sources, and capable of exchanging work with the outside world through processes of expansion and compression.

Carnot's principle, which applies to processes of this kind, is already a practical expression of the second principle; he, in particular, evaluated the efficiency of the cycle, expressed in terms of the amount of work produced and the amount of heat transferred to the system:

$$\eta = 1 - \frac{T_2}{T_1} \tag{1.8}$$

where T_1 and T_2 are, respectively, the temperatures of the hottest and coldest heat source; this efficiency is the maximum obtainable with two sources at T_1 and T_2,

[1] For an infinitesimal displacement $\vec{ds}$, the work of a force $\vec{F}$ is defined as $|\delta\mathscr{L}| = \vec{F} \cdot \vec{ds}$ where $\vec{F}$ is the force exerted on a moving wall of surface area A (e.g., of a piston) by a gas contained in a vessel at pressure P:

$$F = P \cdot A$$

consequently, to a transformation with which a dx displacement of the piston is associated:

$$\delta\mathscr{L} = d\mathscr{L} = -PA\,dx = -PdV.$$

given the reversibility of the processes considered; in addition, the unit value of η holds only in the case of $T_2 = 0\,K$a limit that cannot be physically reached, which also establishes that the efficiency of a thermal machine acting by virtue of heat exchange with two heat sources can never be unitary. From this experience, Clausius found the expression for cycles:

$$\oint \frac{dQ}{T} = 0 \tag{1.9}$$

In other words, the new function in differential terms $dS = \frac{dQ}{T}$ is a state function; this thermodynamic function S is termed *entropy*. Specifically, with reference to an infinitesimal transformation, the associated entropy can be expressed as:

$$dS = d_e S + d_i S \tag{1.10}$$

The term $d_e S$ is referred to matter and energy exchange with the environment; for a closed, mechanically isolated system, $d_e S = \frac{dQ}{T}$; on the other hand, the quantity $d_i S$ is related to the entropy change associated with internal irreversible processes.

The second principle of thermodynamics, in Clausius' formulation, is concerned with defining constraints concerning $d_i S$, and in particular, the following rule generally applies:

$$d_i S \geq 0 \tag{1.11}$$

where equal refers to reversible processes, which do not generate entropy; irreversible processes, therefore, are associated with an increase in the entropy of the system; in other words, a generic infinitesimal process is associated with a change:

$$dS \geq \frac{dQ}{T} \tag{1.12}$$

This expression is also known as *Clausius' inequality*.

This formulation gives rise to a cosmological extension, which is attributed to Clausius himself: the entropy of an isolated system (the entire universe) tends to increase.

The thermodynamic equilibrium proposed by this principle is, for the universe, the heat death, which corresponds to the maximum value of entropy that the nature of real processes would tend to reach, inherently irreversible in nature.

To define heat, we must resort to the second law of thermodynamics, which is based on the definition of a new state function called entropy, of which we have already given an extensive molecular interpretation.

For an infinitesimal and reversible transformation, the entropy S is defined in terms of variations as:

$$dS = \frac{\delta Q}{T} \Longrightarrow \delta Q = TdS \tag{1.13}$$

The change in entropy is zero, therefore, for infinitesimal and reversible transformations of closed systems in which there is no heat exchange (adiabatic). In other words, a reversible adiabatic transformation occurs at constant entropy (isoentropic).

Substituting definition 1.13 for heat and the definition for mechanical work, for closed systems—discontinuous ones—in the absence of chemical reactions, we have:

$$dU = TdS - PdV \tag{1.14}$$

For a closed system containing c components, in which chemical reactions occur, a term is introduced to account for the variations in the number of moles of the c components dn_i:

$$\Delta U_{chim} = \sum_{i=1}^{c} \mu_i dn_i \implies dU = TdS - PdV + \sum_{i=1}^{c} \mu_i dn_i \tag{1.15}$$

This expression, valid for an infinitesimal and reversible transformation, is also valid in the case of continuous systems, with or without chemical reactions, as the variations in moles due to the inflow and outflow currents of the system, together with the effect of chemical reactions, are included in the same term 1.15.

1.3 Thermodynamic Potentials and Chemical Equilibrium

The internal energy U and entropy S are state functions and, for this reason, can be associated with thermodynamic transformations, referring to an initial state A and a final state B.

The characteristic of these state functions, always defined as the difference between the values assumed by the function between two states, actually makes them similar to gravitational potentials, which have no absolute value but are relative to a considered reference value.

For this reason, the term "thermodynamic potentials" is used to refer to state functions in general (particularly U and S), always defined by values relative to a reference state. The reference state generally requires the definition of pressure, temperature, phase (which depends on pressure and temperature), and composition.

1.3.1 Legendre Transformation and Thermodynamic Potentials

Sometimes, thermodynamic potentials are expressed as functions of independent variables that are difficult to handle. For example, if we analyze the general expression for the variation of internal energy U 1.15, we have $U = U(S, V, n_i)$.

Consequently, for example, temperature, pressure, and chemical potential are defined as partial derivatives of this function U:

$$T = \left(\frac{\partial U}{\partial S}\right)_{V,n_i}$$
$$P = -\left(\frac{\partial U}{\partial V}\right)_{S,n_i} \tag{1.16}$$
$$\mu_i = \left(\frac{\partial U}{\partial n_i}\right)_{S,V,n_{j\neq i}}$$

As evident, not all independent variables are equally accessible and convenient. For example, it is difficult to practically achieve a controlled variation of entropy since entropy is a derived property of the system, not directly accessible for measure and control.

Legendre transforms are a simple and effective mathematical tool for changing the dependence of functions on independent variables. This characteristic is widely used in thermodynamics to express the state functions in the transition from one system to another.

The general mathematical problem consists of transforming a function $f(x, y, z)$ into a new function correlated with it, expressing the following functional dependence: $g(u, y, z)$, where $u = \frac{\partial f}{\partial x}$. Furthermore, defining $v = \frac{\partial f}{\partial y}$ and $w = \frac{\partial f}{\partial z}$, the differential of the function $f(x, y, z)$ is given by $df = udx + vdy + wdz$. The new function $g(u, y, z)$ will be defined as $g = f - ux$; indeed, the total differential of g will be $dg = df - udx - xdu = vdy + wdz - xdu$. In this way, the desired transformation $f(x, y, z) \rightarrow g(u, y, z)$ has been achieved, with $x = -\frac{\partial f}{\partial u}$ and $v = \frac{\partial f}{\partial y}$, $w = \frac{\partial f}{\partial z}$. The variables x and u are called *conjugate variables.*

In mathematical language, Legendre transformations are linear variations of variables in which one or more products of conjugate variables are subtracted from the starting function $f(x, y, z)$ to define a new mathematical function $g(u, y, z)$.

In classical thermodynamics, Legendre transformations allow us to derive a series of state functions that depend on the thermodynamic quantities of interest, starting from the internal energy state function $U(N, V, S)$, through the use of Maxwell's relations.

In detail, for a system consisting of a single phase and a single component (with N moles), for an infinitesimal reversible transformation, we can write the differential of the (specific) internal energy as:

$$dU = TdS - pdV + \mu dN \tag{1.17}$$

From this relationship, we obtain $T = \left(\frac{\partial U}{\partial S}\right)_{V,N}$, $p = \left(\frac{\partial U}{\partial V}\right)_{S,N}$, and $\mu = \left(\frac{\partial U}{\partial N}\right)_{S,V}$.

It is evident that we cannot directly measure the variations in entropy associated with a transformation, even if it is reversible. Therefore, there is a need to find new state functions, starting from the internal energy, that have more appropriate characteristics, meaning that their set of independent variables is more accessible to measurements.

From this need, all commonly known thermodynamic quantities and even others of more limited use have been defined.

Firstly, we can choose entropy S as the independent variable and interchange V with its conjugate variable p. The new function is called enthalpy and is defined as:

$$H = U - \left(\frac{\partial U}{\partial V}\right)_{S,N} \cdot V + \mu \cdot N = U + pV + \mu \cdot N \tag{1.18}$$

Expanding the differential of this function, we have $dH = TdS - pdV + \mu dN + pdV + Vdp$. From this, we can indeed obtain $H = H(S, N, p)$, with $T = \left(\frac{\partial H}{\partial S}\right)_{p,N}$, $V = \left(\frac{\partial H}{\partial p}\right)_{S,N}$, and $\mu = \left(\frac{\partial H}{\partial N}\right)_{S,p}$.

Similarly, starting from the definition of internal energy, we can derive a function called the Helmholtz free energy A, which is a function of T, V, and N. It is obtained by interchanging the two conjugate variables T and S, that is, $A = U - TS$. Expanding the total differential, we have $dA = -SdT - pdV + \mu dN$, and thus, $A = A(T, V, N)$.

To obtain a state function that depends not only on the number of moles N but also on P and T, that is, to have a state function that can be determined based on the two more easily controllable parameters, such as pressure and temperature, a double Legendre transformation must be performed, which involves reversing the pairs of conjugate variables $S \rightarrow T$ and $V \rightarrow p$. The resulting function is called Gibbs free energy and is denoted by G. It is defined as $G = U - \left(\frac{\partial U}{\partial V}\right)_{S,N} \cdot p - \left(\frac{\partial U}{\partial S}\right)_{V,N} S = U + pV - TS = H - TS = A + pV$. The differential of G can be obtained as $dG = -SdT + Vdp + \mu dN$.[2]

1.3.2 Partial Molar Properties and Chemical Potential

Let Φ be an extensive thermodynamic quantity evaluated for a thermodynamic system with given T, P, and composition (c components). The corresponding specific quantity ϕ is given by:

$$\varphi = \frac{\Phi}{\sum_{i=1}^{c} n_i} = \frac{\Phi}{n} \tag{1.19}$$

The partial molar quantity of the extensive quantity Φ with respect to the i-th component is given by:

$$\bar{\phi}_i = \left(\frac{\partial \Phi}{\partial n_i}\right)_{T,P,n_{j \neq i}} \tag{1.20}$$

[2] For a double Legendre transformation, from $f = f(x, y)$ to $g = g(u, v)$, we have $dg = -xdu - ydv + wdz$.

Examples of partial molar quantities are the partial molar volume:

$$\bar{v}_i = \left(\frac{\partial V}{\partial n_i}\right)_{T,P,n_{j\neq i}} \tag{1.21}$$

and the chemical potential:

$$\mu_i(T, P) = \left(\frac{\partial U}{\partial n_i}\right)_{S,V,n_{j\neq i}} = \left(\frac{\partial H}{\partial n_i}\right)_{S,P,n_{j\neq i}} = \left(\frac{\partial A}{\partial n_i}\right)_{T,V,n_{j\neq i}}$$
$$= \left(\frac{\partial G}{\partial n_i}\right)_{T,P,n_{j\neq i}} \tag{1.22}$$

Extensive quantities have the property of being homogeneous functions[3] of degree 1, therefore:

$$\Phi(T, P, kx_1, kx_2, \ldots, kx_n) = k \cdot \Phi(x_1, x_2, \ldots, x_n) \tag{1.23}$$

(Euler's theorem)

If a function $f(x_1, x_2, \ldots, x_n)$ is homogeneous of degree m with respect to variables $x_1, x_2, \ldots, x_n$, then

$$\sum_{i=1}^{n} \frac{\partial f}{\partial x_i} \cdot x_i = m \cdot f(\mathbf{x}) \tag{1.24}$$

Applying Euler's theorem to a generic extensive thermodynamic variable, we find:

$$\Phi = \sum_{i=1}^{c} n_i \cdot \bar{\phi}_i \tag{1.25}$$

Since $\Phi = n \cdot \varphi$, we have:

$$\varphi = \sum_{i=1}^{c} x_i \cdot \bar{\phi}_i \tag{1.26}$$

$$\bar{\Phi}_i = \left[\frac{\partial}{\partial n_i}(n \cdot \varphi)\right]_{T,P,n_{j\neq i}} = n \cdot \left(\frac{\partial \varphi}{\partial n_i}\right) + \varphi \cdot \left(\frac{\partial n}{\partial n_i}\right) = \varphi + n \cdot \left(\frac{\partial \varphi}{\partial n_i}\right) \tag{1.27}$$

[3] A function is said to be homogeneous of degree m with respect to variables $\mathbf{x}$ if:

$$\Phi(k\mathbf{x}) = k^m \cdot \Phi(\mathbf{x}).$$

We obtain:

$$\bar{\Phi}_i = \varphi - \sum_{\substack{k=1 \\ k \neq i}}^{c} \left(\frac{\partial \varphi}{\partial x_k}\right)_{T,P,x_{j\neq k}} \cdot x_k \tag{1.28}$$

For a binary system:

$$\begin{cases} \bar{\Phi}_1 = \varphi + (1 - x_1) \cdot \frac{\partial \varphi}{\partial x_1} \\ \bar{\Phi}_2 = \varphi - x_1 \cdot \frac{\partial \varphi}{\partial x_1} \end{cases} \tag{1.29}$$

By applying Euler's theorem, it is possible to derive another important relationship valid for partial molar quantities, which is often used to assess the consistency of experimental data in the field of physical chemistry.

Given the intensive quantity $\Phi\,(P,\ T,\ ,n_1,\ \ldots,\ n_c)$, its differential will be:

$$d\Phi = \left(\frac{\partial \Phi}{\partial P}\right)_{T,n_i} dP + \left(\frac{\partial \Phi}{\partial T}\right)_{T,n_i} dT + \sum_{i=1}^{c} \bar{\Phi}_i \cdot n_i \tag{1.30}$$

On the other hand, according to Euler's theorem,

$$\Phi = \sum_{i=1}^{c} n_i \cdot \bar{\Phi}_i \tag{1.31}$$

whose differential is:

$$d\Phi = \sum_{i=1}^{c} n_i \cdot d\bar{\Phi}_i + \sum_{i=1}^{c} \bar{\Phi}_i \cdot dn_i \tag{1.32}$$

It is then possible to get the following expression:

$$\sum_{i=1}^{c} n_i \cdot d\bar{\Phi}_i = \left(\frac{\partial \Phi}{\partial P}\right)_{T,n_j} + \left(\frac{\partial \Phi}{\partial T}\right)_{P,n_j} \tag{1.33}$$

In terms of molar fractions, it is rewritten as:

$$\sum_{i=1}^{c} x_i \cdot d\bar{\Phi}_i = \left(\frac{\partial \phi}{\partial P}\right)_{T,x_j} + \left(\frac{\partial \phi}{\partial T}\right)_{P,x_j} \tag{1.34}$$

This equation is known as the *Gibbs-Duhem equation*; a T and P, constant, Eq. 1.35 becomes:

$$\sum_{i=1}^{c} x_i \cdot d\bar{\Phi}_i = 0 \tag{1.35}$$

For binary systems, it becomes:

$$\frac{\partial \bar{\Phi}_2}{\partial x_1} = -\frac{x_1}{x_2} \cdot \frac{\partial \bar{\Phi}_1}{\partial x_1} \tag{1.36}$$

This expression is particularly important since it is used to test the consistency of experimental data: for instance, if $\frac{\partial \bar{\Phi}_2}{\partial x_1}$ is positive, $\frac{\partial \bar{\Phi}_1}{\partial x_1}$ should be negative or vice versa.

1.3.3 Fugacity, Chemical Potential, and Activity

The chemical potential is defined, in a very general way, as the partial molar quantity of thermodynamic potentials; specifically, for open systems, the chemical potential of a generic component i is defined in terms of Gibbs free energy G as:

$$\mu_i = \left(\frac{\partial G}{\partial n_i}\right)_{T,P,n_{j \neq i}} \tag{1.37}$$

Given a monocomponent system, at a given pressure P and temperature T, in general, it is possible to define the corresponding specific molar Gibbs free energy g:[4]

$$dg\,(T,P) = v dP - s\,dT \tag{1.38}$$

[4] Given an extensive thermodynamic property Φ for a system containing n moles, the corresponding specific molar quantity is defined as:

$$\varphi = \frac{\Phi}{n}$$

being an average value with respect to the *total* concentration. The corresponding partial molar quantity $\bar{\Phi}_i$, on the other hand, depends on single components' concentration, such as:

$$\bar{\Phi}_i = \left(\frac{\partial \Phi}{\partial c_i}\right)_{j \neq i}$$

These two quantities are linked as:

$$\bar{\Phi}_i = \varphi - \sum_{\substack{k=1 \\ k \neq i}}^{c} \left(\frac{\partial \varphi}{\partial x_K}\right)_{T,P,,x} \cdot x_K$$

For a binary system:

$$\bar{\Phi}_1 = \varphi + (1 - x_1) \cdot \frac{\partial \varphi}{\partial x_1}$$

$$\bar{\Phi}_2 = \varphi - x_1 \cdot \frac{\partial \varphi}{\partial x_1}.$$

as a function of pressure, temperature, and molar entropy s and specific volume v; based on this definition, it is possible to write:

$$v = \left(\frac{\partial g}{\partial P}\right) T \qquad s = \left(\frac{\partial g}{\partial T}\right) P \tag{1.39}$$

The definition of v introduces molecular parameters through the equation of state.

Considering an isothermal transformation of a system, in general, it is possible to write:

$$dg_T = v dP \tag{1.40}$$

for a pure, perfect gas, $v = \frac{RT}{P}$, then:

$$dg_T = \frac{RT}{P} dP = RT \cdot d \log P \tag{1.41}$$

that integrated between the two pressures P_1 to P_2, produces the corresponding variation of specific Gibbs free energy:

$$\Delta g_T = RT \cdot \log \frac{P_2}{P_1} \tag{1.42}$$

Generalizing Eq. 1.41 for a real gas, it is possible to define a new thermodynamic quantity, the *fugacity*, defined as:

$$\left\{ d\mu = RT, \log f \ \lim_{P \to 0} \frac{f}{P} = 1 \right. \tag{1.43}$$

From this, taking into account that, under isothermal and isobaric conditions, that $d\mu = dG\,(T, P)$, it is possible to write:

$$d\mu = RT \cdot d \log f \tag{1.44}$$

Considering a reference state, denoted by the symbol $\oplus$, the chemical potential at a given T and P is:

$$\mu(T, P) = \mu^{\oplus} + RT \cdot \log \frac{f(T, P)}{f^{\oplus}} \tag{1.45}$$

Considering as a reference phase at $P^* \to 0$ (perfect gas), indicating by "*" the perfect gas reference state, the fugacity of a pure perfect gas being $f^* = P^*$, the corresponding chemical potential is:

$$\mu(T, P) = \mu^*(T, P^*) + RT \cdot \log \frac{f(T, P)}{P} \tag{1.46}$$

It is possible to derive from 1.46 the following expression:

$$\log \frac{f}{P} = \log \nu = \frac{1}{RT} \cdot \int_0^P \left(v - \frac{RT}{P} \right) \tag{1.47}$$

where ν is the *fugacity coefficient*. For condensed phase, in Eq. 1.45, the ratio $\frac{f(T,P)}{f^{\oplus}}$ is defined as a new property, called *activity*, that is defined for the component i as:

$$a_i = \frac{f(T, P)}{f^{\oplus}} = \gamma_i \cdot x_i \tag{1.48}$$

γ_i is the *activity coefficient*, which depends on the concentration of all components and temperature, while x_i is the molar fraction of the component i. Applying this definition, the chemical potential of the component i in a condensed phase is described by the following expression:

$$\mu_i(T, x_i) = \mu_i^0 + RT \cdot \log a_i \tag{1.49}$$

where the reference chemical potential μ_i^0 is not necessarily set at the same condensed phase.

1.3.4 Thermodynamic Phase Equilibria

The general definition of the thermodynamic equilibrium is based on the presence of sole reversible processes ($d_i S = 0$) and states that for a system at T and P constant, the variation of the Gibbs free energy should be zero:

$$dG = 0 \tag{1.50}$$

As previously defined, the variation of G is expressed as:

$$dG = -SdT - VdP + \sum_{i=1}^{c} \mu_i dn_i \tag{1.51}$$

so the thermodynamic equilibrium at fixed T and P declines as:

$$\sum_{i=1}^{c} \mu_i = 0 \tag{1.52}$$

Considering the equilibrium between a phase α and β, the equilibrium is given by:

$$\mu_i^{\alpha} = \mu_i^{\beta} \tag{1.53}$$

If the reference states are the same for the chemical potential in the two phases, Eq. 1.53 becomes:

$$f_i^{\alpha} = f_i^{\beta} \tag{1.54}$$

For this reason, the condition of thermodynamic condition of a component between two phases is referred to as *isofugacity condition*.

Similarly, for condensed phases, Eq. 1.53 in case of the same reference state for both phases can be written as:

$$a_i^{\alpha} = a_i^{\beta} \tag{1.55}$$

and referred to as *isoactivity condition*.

Additional Resources and Recommended Literature

The book [18] is a general textbook on chemical thermodynamics, from theory and definitions to applications. The book [16] is a classic textbook on the application of molecular theories to the understanding of chemical thermodynamics.

2 Statistical Thermodynamics

Life is like a box of chocolates. You never know what you're gonna get. - Forrest Gump.

E. Roth

Abstract

Why it is important to know this material?

Statistics provides the key elements for understanding the molecular nature of the macroscopic properties of thermodynamic systems. It provides key elements for understanding.

What is the key idea?

The central idea underlying statistical mechanics and thermodynamics is that the macroscopic properties can be obtained by averaging the molecular properties of the conformation group, i.e. the set of all possible molecular conformations based on the physical constraints of thermodynamic systems (such as constant pressure or temperature).

What is necessary to know already?

The basis of the statistics must be very well set, such as the knowledge of the basic relationship between systems properties and thermodynamic potentials.

2.1 Introductive Elements

Classical thermodynamics deals with the formulation of relationships among different properties of macroscopic systems (i.e., containing more than 10^{20} molecules); the laws of thermodynamics are based on the definition of thermodynamic systems, models of real systems with given properties.

L. Di Paola, *Fundamentals of Molecular Bioengineering*,
https://doi.org/10.1007/978-3-031-42022-1_2

The classification into thermodynamic systems allows for the application of general relationships to real systems with analogous characteristics, and through these relationships, the value of quantities that are not directly accessible to experimental measurement can be derived from quantities that are directly and easily measurable.

However, these laws do not directly describe the microscopic behavior of the system. Nonetheless, macroscopic quantities are derived from the average over the entire number of molecules or atoms constituting the macroscopic systems, of correlated microscopic quantities.

Statistical thermodynamics primarily deals with precisely this: by making certain assumptions and starting from specific postulates, it provides the physico-statistical tools to derive some macroscopic properties from molecular quantities.

To achieve this objective, statistical thermodynamics makes use of general tools from statistical mechanics to translate the mechanical properties of individual molecules that constitute the system into macroscopic properties.

The ensemble method provides an indispensable for defining the statistical methods necessary to transition from the microscopic scale—molecular or submolecular—to the macroscopic scale.

Statistical thermodynamics also produces theorems that cannot be deduced exclusively using classical thermodynamics; these tools enable the evaluation of macroscopic quantities from molecular information (e.g., the method for calculating heat capacities from spectroscopic data).

Finally, through quantum mechanics, it is possible to describe the energy distribution on the molecular scale by defining quantum mechanical constraints regarding the spatial and mechanical configurations of the set of molecules that constitute the system (*states*), distinguishing between accessible and inaccessible states.

2.2 Kinetic Theory of Gases

The kinetic theory of gases was the first attempt to establish a relationship between the microscopic properties of a physical system and its corresponding macroscopic characteristics. In particular, through a purely representational approach, this theory is based on the concept that all properties of matter can be explained in terms of the motion properties of the individual constituents of matter (atoms, molecules, or supermolecular aggregates).

The kinetic theory was originally developed to describe the properties of gases at low pressure and relatively high temperatures, which could be macroscopically described by the ideal gas law, as will be extensively discussed in the following paragraph.

2.2.1 Molecular Interpretation of a Gas Pressure and Relationship with Internal Energy

Let's consider N moles of gas inside a container at constant volume and temperature, the pressure in the container is function of N, the volume of the container V, and the temperature T:

$$P = P(N, V, T) \tag{2.1}$$

This relationship is known as the *equation of state* , and its specific form varies depending on the molecular characteristics of the gas, particularly the intermolecular interactions, as extensively illustrated in the following chapters.

However, pressure also represents the ratio between the force applied to a surface and the surface area itself:

$$P = \frac{F}{A} \tag{2.2}$$

In other words, the pressure exerted on the surfaces of a container containing the gas is the result of a force, which arises from the collisions of the molecules against those surfaces.

In fact, at the molecular and atomic level, even an object that appears to be stable and unchanging macroscopically is composed of components that are in constant and chaotic motion, which determine its properties. Thus, even though the gas container macroscopically exhibits the same characteristics of pressure, number of moles, volume, and temperature, the microscopic reality that comprises it consists of a multitude of molecules or atoms in continuous, uninterrupted motion.

The properties associated with the motion of molecules, averaged over the entire number of molecules, are translated into macroscopic quantities that can be measured using common instruments (such as a thermometer or a barometer).

Regarding pressure,[1] for instance, it is given by:

$$P = \frac{2}{3} \cdot n \cdot \left\langle m \cdot \frac{v^2}{2} \right\rangle = \frac{2}{3} \cdot \frac{N}{V} \mathscr{N} \cdot \left\langle m \cdot \frac{v^2}{2} \right\rangle \tag{2.3}$$

where $\left\langle m \cdot \frac{v^2}{2} \right\rangle$ represents the kinetic energy of the center of mass of the particle system composed of all the molecules,[2] n is the number of molecules per unit volume, and $\mathscr{N}$ is Avogadro's number.

[1] For a complete treatment of the problem, reference should be made to a textbook on General Physics.

[2] As known, the kinetic energy of the center of mass is also equal to the average kinetic energy of the individual molecules, considering a reference set centered to the center of mass of single molecules.

This relationship is derived under the assumption of energy transfer through completely elastic collisions, meaning that molecules are considered infinitely rigid objects that do not deform upon collision. In reality, molecules do deform due to collisions, and part of the energy is therefore dissipated in this deformation.

For a monoatomic gas, the internal energy is given by the kinetic energy of the molecules:

$$U = n \cdot \left\langle m \cdot \frac{v^2}{2} \right\rangle \rightarrow PV = \frac{2}{3} \cdot U \tag{2.4}$$

This expression, valid for monoatomic gases, can be generalized for polyatomic gases:[3]

$$PV = (\gamma - 1) \cdot U \tag{2.5}$$

where, compared to Eq. 2.4, $\gamma = \frac{5}{3}$ is obtained for a monoatomic gas.

2.2.2 Relationship Between Temperature and Molecular Properties

Starting from the general expression 2.4, which relates the internal energy of a gas to pressure, and assuming that the internal energy of the gas is given in terms of the average kinetic energy of the molecules, it is possible to get:[4]

$$PV = (\gamma - 1) \cdot N\mathcal{N} \cdot V \cdot \left\langle m \cdot \frac{v^2}{2} \right\rangle \tag{2.6}$$

For a gas at low pressure and high temperature, the ideal gas law can be used:

$$PV = NRT = (\gamma - 1) \cdot N\mathcal{N} \cdot \left\langle m \cdot \frac{v^2}{2} \right\rangle \tag{2.7}$$

From this relationship, the relation between temperature and molecular kinetic energy is derived:

$$\left\langle m \cdot \frac{v^2}{2} \right\rangle = \frac{1}{\gamma - 1} \cdot k_B \cdot T \tag{2.8}$$

[3] Even in the case of polyatomic gases, it is assumed that the individual molecules are spherical and infinitely rigid. It is evident that in this case, more so than for monoatomic gases, these assumptions are less representative of the microscopic reality.

[4] In the case of polyatomic gases, there is rigorously an additional contribution due to intramolecular motions, which are considered negligible in this treatment; this additional assumption is actually a direct consequence of the first assumption of infinite rigidity of the molecule.

where the constant $k_B = \frac{R}{\mathcal{N}} = 1.38 \cdot 10^{-23}$ J/K is called the Boltzmann constant and is one of the universal constants widely used in statistical and molecular thermodynamics.

For a monoatomic gas, $\frac{1}{\gamma - 1} = \frac{3}{2}$, thus:

$$\left\langle m \cdot \frac{v^2}{2} \right\rangle = \frac{3}{2} k_B T \tag{2.9}$$

It should be noted that $\left\langle m \cdot \frac{v^2}{2} \right\rangle$ represents the kinetic energy of the center of mass of the molecules in three-dimensional space, so $\frac{3}{2} k_B T$ represents the kinetic energy for motion in three directions, from which it follows that the contribution to motion in a single direction is $\frac{1}{2} k_B T$.

2.3 Principles of Statistical Mechanics

As seen above, the kinetic theory of gases is based on a mechanistic interpretation of the macroscopic properties of matter, which can be derived from the motion characteristics of the individual constituent particles (atoms or molecules).

In the previous paragraph, it was explained how for systems in thermal equilibrium, it is possible to derive the macroscopic properties of the system (density, pressure, and temperature) by averaging the mechanical properties of the microscopic components. This type of approach generally falls within the characteristic methods of statistical mechanics.

To perform averages of the mechanical properties over the entire particle system, it is necessary to know the statistical distributions of these quantities (particle positions and velocities). For example, consider a column of single-component gas with a very large height in thermal equilibrium.[5] The spatial distribution of gas molecules per unit volume n varies with the height of the column h according to the law:

$$n = n_0 \cdot \exp\left(-\frac{m\,g\,h}{k_B\,T}\right) \tag{2.10}$$

where n_0 is the density at height $h = 0$, m is the molecular weight of the gas molecules, and g is the acceleration due to gravity.

By comparing data for different gases, Eq. 2.10 reveals that the higher the molecular weight, the more rapidly the gas becomes rarified. This would indicate, in a gas mixture, an enrichment of lighter gases in the upper layers.

[5] This assumption is highly restrictive since the Earth's atmosphere is intrinsically not in thermal equilibrium: the temperature decreases with increasing altitude.

The density distribution 2.10 is generalized in the form known as Boltzmann's law:

$$n = n_0 \cdot \exp\left(-\frac{w}{k_B\,T}\right) \tag{2.11}$$

where $w = m\,g\,h$ is the potential energy acting on each individual atom or molecule that constitutes the gas. This law, derived in the case of a gravitational field, can be extended to different cases involving other forces. The law 2.11 describes an exponential decay of the reference value n_0 for the distribution. In particular, the term $\exp\left(-\frac{w}{k_B\,T}\right)$ is called the *Boltzmann factor*, and the argument of the exponential is the ratio of the potential energy w associated with the applied force field to the term $k_B\,T$, which represents the energy due to thermal agitation, equivalent to the average kinetic energy of the molecules, which varies proportionally with temperature (Eq. 2.9) (Fig. 2.1).

Boltzmann's law represents, in its most generalized form, a fundamental principle of statistical mechanics:

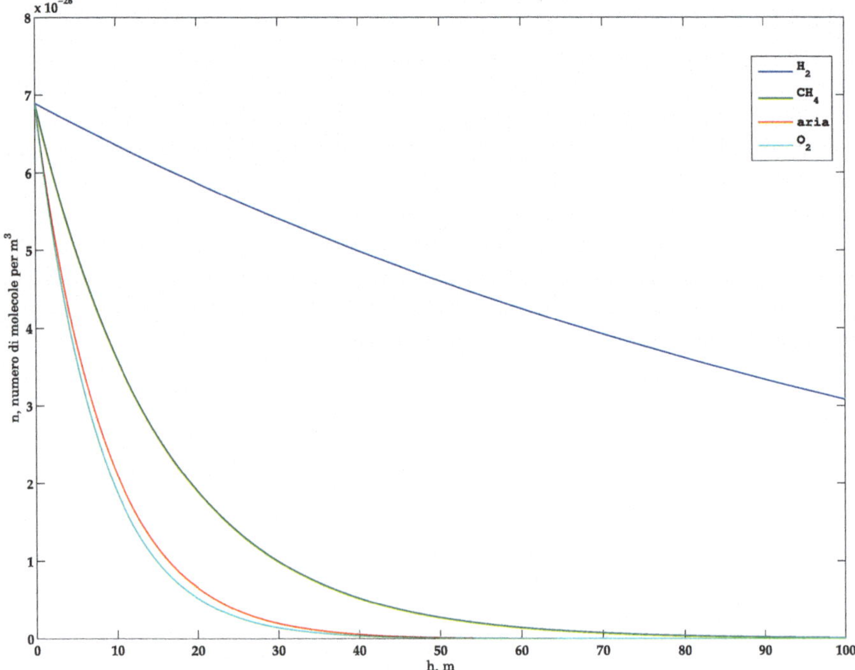

Fig. 2.1 Density distribution for some gases, at 20°C

The probability of finding molecules in a particular spatial distribution varies proportionally with the negative exponential of the ratio between potential energy and thermal activation energy.

In this sense, the potential energy w represents the molecular order, coordinated by the intermolecular interactions, which opposes the term of chaotic agitation $k_B T$, which tends to disrupt the order imposed on matter by the lines of applied intermolecular forces.

If $w \gg k_B T$, the potential energy prevails over thermal agitation, resulting in a strong effect of the order imposed by the potential energy, leading to rapid rarefaction. On the other hand, if $w \ll k_B T$, the disorder effect imposed by strong thermal agitation prevails, and the density is nearly uniform along the vertical axis.

2.3.1 Distribution of Molecular Velocities

One of the assumptions underlying the law 2.11 is that the gas can be considered ideal (ideal gas equation of state): in other words, the interactions between gas molecules can be neglected.

In this case, assuming the absence of collisions as well, it is possible to estimate the corresponding distribution of molecular velocities.

Molecules at $h = 0$ will have different initial velocities, allowing different molecules to reach different heights. Considering the vertical component of velocity v_z of a generic molecule at $h = 0$, at each height h, we can associate a characteristic velocity u such that only if $v_z \geq u$ (minimum value), the molecule can reach the height h (trajectory 2, Fig. 2.2); this minimum value is found by applying the principle of conservation of mechanical energy, which states:

$$\frac{1}{2} m\, u^2 = m\, g\, h \;\rightarrow\; u = \sqrt{2\, g\, h} \tag{2.12}$$

Under isothermal conditions, the density distribution will be the same for all particles. Therefore, the ratio between the number of molecules n_2 with $v_z \geq u$ (trajectory 2, Fig. 2.2) at $h = 0$ and the number n_1 of molecules with $v_z \geq 0$ ($u = 0$ for $h = 0$, trajectory 1, Fig. 2.2) will be:

$$\frac{n_2}{n_1} = \frac{n_0 \cdot \exp\left(-\frac{m\,g\,h}{k_B T}\right)}{n_0} = \exp\left(-\frac{m\,g\,h}{k_B T}\right) = \exp\left(-\frac{m\,u^2}{2\,k_B T}\right) \tag{2.13}$$

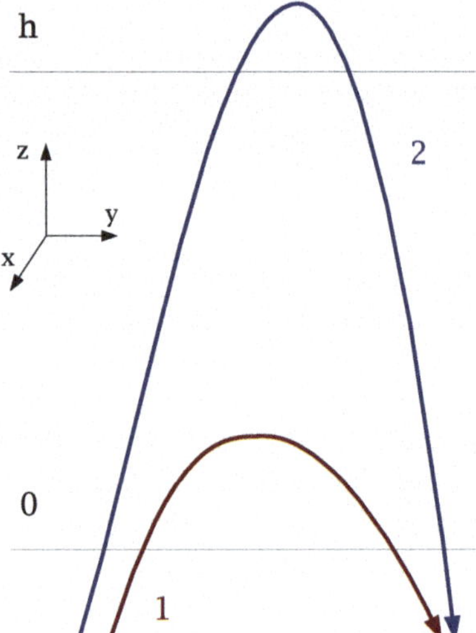

Fig. 2.2 Velocities' distribution: molecules' trajectories

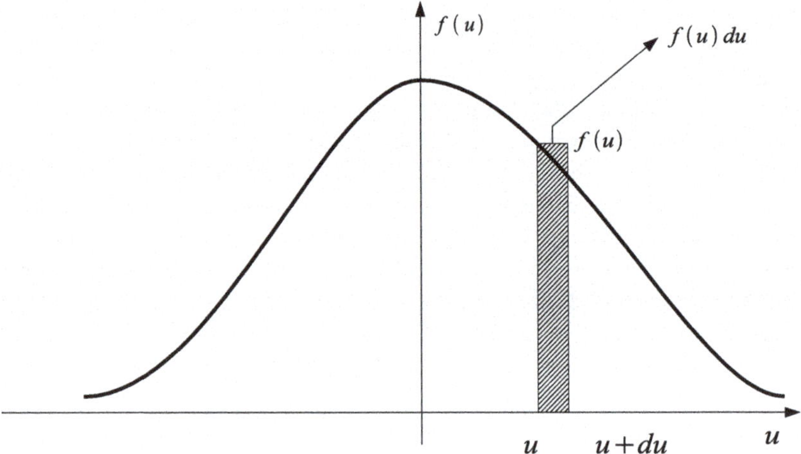

Fig. 2.3 Distribution of molecular velocities

This result can be generalized for any type of molecular velocity distribution:

$$n \propto e^{-\frac{\text{kinetic energy}}{k_B T}} \tag{2.14}$$

The velocity distribution can be expressed through a continuous function $f\,(u)$ as shown in Fig. 2.3. The area enclosed by the rectangle shown in the figure represents the fraction of molecules that have a velocity between u and $u+du$. The distribution must satisfy the condition $\int_0^{+\infty} f\,(u)\,du = 1$.

The number of molecules that pass through a certain section with a velocity greater than a certain value u is given by:

$$\int_{u}^{+\infty} u\, f(u)\, du = \sqrt{\frac{m}{2k_B T}} \cdot e^{-\frac{mu^2}{2k_B T}} \tag{2.15}$$

This expression is derived by considering that faster molecules pass through the same section with a higher frequency, so the average is calculated using the velocities u of the molecules as weights.

2.3.2 Application of Kinetic Theory of Gases

Briefly summarizing, the main results of the kinetic theory of gases are as follows:

1. The contribution of the kinetic energy corresponding to each degree of freedom is equal to $\frac{1}{2} k_B T$.
2. The probability of finding a particle in a specific spatial position is proportional to $e^{-\frac{\text{potential energy}}{k_B T}}$.
3. The probability that a particle moves with a certain velocity is proportional to $e^{-\frac{\text{kinetic energy}}{k_B T}}$.

These laws are not suitable for the rigorous description of systems containing a large number of interacting particles. However, under strong simplifying assumptions, it is possible to evaluate some chemical and physical properties of macroscopic systems, as will be illustrated below.

2.3.2.1 Evaporation

The first example of applying the principles of statistical mechanics concerns the evaporation of a liquid. Consider a large-volume container partially filled with a liquid in equilibrium with its corresponding vapor at a given temperature. It is assumed that the vapor molecules are much farther apart from each other, while the molecules in the liquid are much closer together, held together by attractive interactions.

Using the tools of statistical mechanics, it is possible to estimate the number n_v of molecules present in the vapor phase compared to those in the liquid phase n_L.[6] n_V and n_L can also be interpreted as the probability of finding a molecule, under given pressure and temperature conditions, respectively in the vapor or liquid phase.

The density distribution can be directly interpreted in terms of a *probability distribution*, where the permissible states are related to the two present phases. This

[6] The distribution n_V varies with temperature T. It is intuitive that by providing heat, i.e., increasing the temperature, evaporation becomes more pronounced, and accordingly, the ratio $\frac{n_v}{n_L}$ increases.

type of interpretation is possible in the perspective of a dynamic molecular reality where, moment by moment, the probability of a molecule belonging to one of the two phases is evaluated. The condition of thermodynamic equilibrium establishes the invariance of macroscopic properties, in other words, the corresponding average values of the particle system, which exhibits microscopically nonuniform and invariant distributions.

Let $\frac{1}{V_a}$ be the number of atoms per unit volume in the liquid, where V_a is the volume occupied by a single molecule in the liquid. The liquid molecules are held together by attractive forces, which are not present between molecules in the vapor phase.

Let W be the work required to bring a single liquid molecule into the vapor phase, representing the difference between the energy possessed by a single molecule in the vapor phase E_V and the corresponding value in the liquid phase E_L:

$$\frac{n_V}{n_L} = e^{-\frac{(E_V - E_L)}{K_B T}} = e^{-\frac{W}{K_B T}} \tag{2.16}$$

The energy difference $E_V - E_L$ between the two states has a direct macroscopic counterpart in the difference in molar specific volumes of the two phases, v_V and v_L, as it is clear in Fig. 2.4; the higher the value of the work W, the greater the difference in density between the two phases.

As the critical point is approached, the densities become more similar, and less work is required to bring the molecules from the liquid to the vapor phase. In fact, as the temperature increases (in the subcritical region of Fig. 2.4 with $T \leq T_C$), the

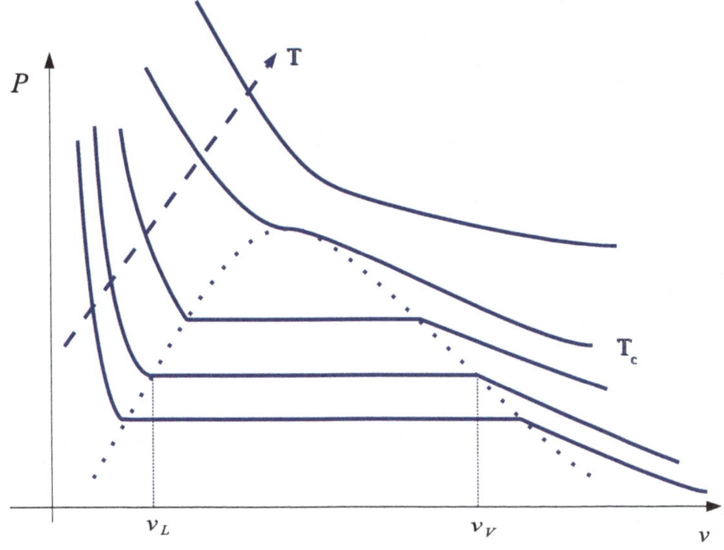

Fig. 2.4 Plane of Clapeyron: isotherms as function of the pressure P and of the molar volume v

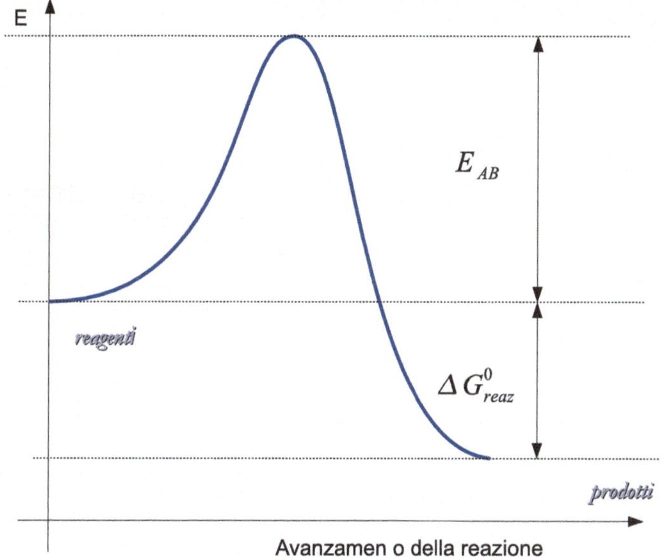

Fig. 2.5 Energy associated with the chemical reaction. $A + B \rightleftharpoons AB^* \rightarrow C$

density difference $\Delta v = v_V - v_L$ decreases, meaning that the work W decreases. In the supercritical region ($T > T_C$), the work $W \ll k_B T$, and only the gas phase can be found, as it cannot be liquefied by compression.

Chemical kinetics deals with evaluating the characteristic times for two molecules A and B to combine into a new compound C. According to a theory called the activated complex theory, this type of reaction follows the following reaction scheme:

$$A + B \rightleftharpoons AB \rightarrow C \tag{2.17}$$

This reaction describes the formation of an activated complex AB, which is not stable and rapidly leads to the formation of products.

The formation of the activated complex AB is conditioned by the collision of molecules A and B. In particular, the collision must be sufficiently violent for the formed complex AB to be resistant to disintegration.

The minimum energy E_{AB} with which A and B must collide to form AB^* is called the activation energy because it is the energy required to activate the reaction itself (Fig. 2.5).

By considering a surface σ_{AB} of contact between molecules of A and B, the problem reduces to determining the fraction of molecules that collide at the surface with a velocity sufficient to initiate the chemical reaction (Fig. 2.6).

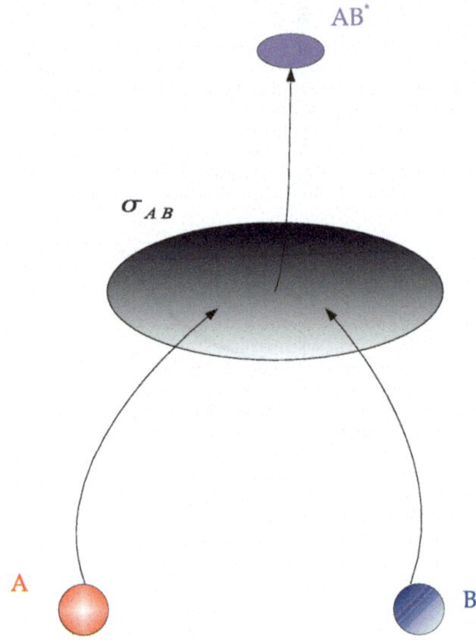

Fig. 2.6 Chemical reactions result from molecular collisions

This problem is equivalent to finding the pairs of molecules A and B that possess sufficient energy E_{AB}, which can be obtained as:

$$R_f = n_A n_B \cdot \sigma_{AB} \, v \, e^{-\frac{E_{AB}}{k_B T}} \tag{2.18}$$

Here, n_A and n_B represent the number of molecules of A and B per unit volume, respectively. The velocity R_f at which these molecules collide has units of $[L^{-3} T^{-1}]$, indicating the number of collisions that occur per unit volume and time. It is worth noting that the collision velocity, which also corresponds to the velocity of the direct reaction $A + B \rightarrow C$, depends on the concentrations n_A and n_B as well as a negative exponential factor $e^{-\frac{E_{AB}}{k_B T}}$. This exponential factor, known as the rate constant, depends only on temperature through the thermal activation energy $k_B T$.

The rate of the reverse reaction $AB^* \rightleftharpoons A + B$ is then given by:

$$R_r = c' \cdot n_{AB} \cdot e^{-\frac{\left(E_{AB} + \Delta G^0_{reaz}\right)}{k_B T}} \tag{2.19}$$

Here, the constant c' incorporates all contributions related to atomic volume and collision velocity. In this case, the energy for the formation of reactants from the activated complex is determined by the sum of the energy required to form the activated complex from the products and the term related to the specific standard free energy difference ΔG^0_{reaz} associated with the reaction.

At equilibrium, these two rates are equal, leading to:

$$R_f = R_r \quad \rightarrow \quad n_A n_B \cdot \sigma_{AB}\, v\, e^{-\frac{E_{AB}}{k_B T}} = c' \cdot n_{AB} \cdot e^{-\frac{\left(E_{AB}+\Delta G^0_{reaz}\right)}{k_B T}} \tag{2.20}$$

From which we obtain:

$$\frac{n_{AB}}{n_A\, n_B} = \mathbb{C} \cdot e^{-\frac{\Delta G^0_{reaz}}{k_B T}} \tag{2.21}$$

Here, $\mathbb{C}$ is a constant that incorporates the cross-sectional area σ_{AB}, molecular velocities, and all other factors independent of E_{AB} and the concentrations n. Equation 2.21 has a similar form to the van't Hoff equation, which relates the equilibrium constant to the change in standard Gibbs free energy ΔG^0_{reaz} associated with the reaction. In other words, the van't Hoff isobar can be derived using methods of statistical mechanics, in the form of Eq. 2.21. In this case, the constants represent specific molecular quantities, but the concept of free energy associated with the reaction retains the same meaning as in the macroscopic derivation law.

2.4 Ensembles and Postulates in Statistical Thermodynamics

In the previous paragraphs, it has been illustrated how the properties of a macroscopic system—pressure and temperature—can be derived in terms of the averages of the mechanical properties of individual particles. In theory, if it were possible to know the mechanical characteristics of the particles—velocity, position, and interactions—instant by instant, it would be possible to derive all the corresponding properties of the macroscopic system using the laws of classical mechanics.

However, this approach, if rigorously applied, is not feasible for macroscopic systems due to the immense number of calculations required and the impossibility of defining the interactions of more than two bodies through an exact method.

In the early 1900s, J.W. Gibbs proposed an alternative procedure to relate macroscopic quantities to microscopic properties of systems, which relied on statistical tools. This method, known as *Gibbs ensemble theory*, is based on the definition of an ensemble of systems and certain postulates concerning the (averaged) properties of these ensembles.

An *ensemble* is defined as a mental collection of a very large number N_S of distinguishable microscopic systems, each of which exhibits the same values of the macroscopic properties as the observed system. For example, let's assume that the system of interest has a volume V, contains N molecules of a single component, and is immersed in a thermal bath at temperature T; from the thermodynamic point of view, the state is completely defined.

The ensemble of interest, used to study the given macroscopic system, consists of an extremely large number N_S of microscopically distinguishable systems, each of which can macroscopically replicate the thermodynamic state (N, V, and T) of the original system. In other words, all the systems in the ensemble are distinct at the microscopic level but indistinguishable from a macroscopic perspective.

The number of possible systems capable of realizing equivalent macroscopic thermodynamic characteristics is very high, mainly because the constraints to be satisfied (macroscopic quantities, N, T, and V) are much fewer compared to the number of particles and, thus, the degrees of freedom of the system at the microscopic level (which is six times the number of particles in the system).

The term *quantum state* of a system refers generally to the overall energy state of the particle system under study, where the total energy is given by the sum of the energies of the individual molecules. This energy cannot vary continuously but takes discrete, quantized values as prescribed by quantum mechanics. However, this view is simplified, and further clarifications are necessary.

The quantum state of an individual molecule is determined by solving the *Schrödinger equation*, which can be approximately derived only in the case of independent molecules (where the intermolecular potential is identically zero throughout space), a situation that occurs only in the case of ideal gases. Using a phrase from Schrödinger himself, the ideal gas is the only system for which it is possible to assign each molecule its "private" energy.

However, similar considerations that lead to the formulation of wave functions for polyatomic molecules allow finding a solution for defining the quantum state of macroscopic systems composed of interacting particles. In this case, the quantum state cannot be derived as a simple sum over the entire particle ensemble but requires following a "systemic" approach, where the entire system is irreducible to its constituent parts due to the strong intermolecular quantized potential.

Another important issue concerns the definition of thermodynamic equilibrium state: a system is in equilibrium if its macroscopic properties remain constant over time. However, at the microscopic level, the state of equilibrium must necessarily be interpreted dynamically, meaning that the corresponding macroscopic values are averages of molecular property values calculated over the entire ensemble and further averaged over the entire time interval required to collect the measurement (which is often several orders of magnitude larger than the typical fluctuations of the system from one quantum state to another).

In these terms, the quantum states of a macroscopic system can be classified as follows:

1. *Accessible states*: These are states that can be easily reached by the system and are reached a large number of times during the measurement interval on the macroscopic system.
2. *Inaccessible states*: These are states that can never be reached in the given time and conditions of the measurement.

The average is always taken over all accessible quantum states, taking into account that they are extremely numerous and rapidly change during a single measurement.

The method of ensembles relies on two postulates, which are unprovable laws that assume the validity of fundamental principles on which the derived results are based.

First Postulate The time average of a mechanical variable M of the thermodynamic system under consideration is equal to the ensemble average of the same variable M, assuming that the systems in the ensemble replicate the thermodynamic state and the environment of the macroscopic system under examination. In other words, the first postulate states that the time average of a quantity M of the system can be replaced by an instantaneous average over a large number of systems, which are equivalent in terms of macroscopic properties to the system under examination.

To state the second postulate, it is necessary to indicate the systems of major interest in the context of equilibrium statistical thermodynamics:

1. *Isolated system* (fixed N, V, and E), corresponding to the *microcanonical ensemble*
2. *Closed system*, isothermal (given N, T, and V), corresponding to the *canonical ensemble*
3. *Open system*, isothermal (given μ, V, and T), corresponding to the *grand canonical ensemble*

Second Postulate In an ensemble ($N_S \rightarrow \infty$) representative of an isolated thermodynamic system (microcanonical ensemble), the systems in the ensemble are uniformly distributed, i.e., with equal probability or frequency, among all possible quantum states consistent with the specific values of N, V, and E. In other words, each quantum state is equiprobable, meaning it is represented by the same number of systems in the ensemble.

When we combine the first and second postulates, we obtain that in the microcanonical ensemble, a single system spends the same amount of time (considering a long time interval) in each quantum state (ergodic hypothesis).

The term *degeneracy* is used to describe the condition where two or more stationary states of the same system have the same energy, even though their wave functions may be different.

In the case of a system with a large number of particles, the value of E can vary almost continuously. Additionally, each energy level (corresponding to a specific value of E) has a high degeneracy, meaning that a total energy level of the system can be realized by numerous different energy configurations of the constituent particles. In general, $\omega(N, V, E)$ represents the number of quantum states associated with an energy level E for a quantum mechanical system composed of N molecules in a volume V.

The two fundamental postulates of statistical thermodynamics establish the characteristic properties of the microcanonical ensemble. From these, the properties of the other two ensembles, canonical and grand canonical, can be derived, as follows.

2.4.1 Canonical Ensemble

The canonical ensemble represents a real system with a fixed volume V and a fixed number of molecules N, placed in a thermostat at a constant temperature T.

In this case, the system is closed but not isolated, so the system's energy fluctuates. The quantum states in the corresponding ensemble are characterized by different energy levels E.

Since mechanical variables have well-defined values in a given quantum state, it remains to determine the fraction of systems in the ensemble that are in a given quantum state (or, equivalently, the probability that a randomly selected system from the ensemble is in a specific quantum state).

By utilizing the system's property of being isothermal (constant temperature T), the N_S systems that make up the ensemble, each with the same N and V, are considered as arranged in a lattice. The dividing walls between the different systems are impermeable to the passage of molecules but not to heat or energy (closed systems).

To ensure that the temperature T is the same for all systems in the lattice, it is ideally placed in a large thermal reservoir at a constant temperature T. At equilibrium, the entire lattice is thermally isolated, and the ensemble is removed from the thermal reservoir. The entire ensemble now becomes a new system (referred to as the *supersystem*) with $N_S N$ molecules, volume $N_S V$, and total energy E_t. Each individual system is placed in a thermal reservoir at temperature T and thus adequately represents the original macroscopic system.

By the way it was constructed, the entire canonical ensemble (the supersystem) is equivalent to an isolated system (microcanonical ensemble), to which the second postulate can be applied to derive the probability of a given quantum state for the systems comprising the canonical ensemble.

Each individual system in the canonical ensemble has the same N and V and, therefore, the same set of accessible quantum states $E_1, E_2, \ldots, E_j, \ldots$.

By observing the entire ensemble at a given instant, it is possible to evaluate the distribution n_j of the number of systems in the quantum state E_j. The conditions that must be satisfied are:

$$\sum_j n_j = N_S \tag{2.22}$$

$$\sum_j n_j E_j = E_t \tag{2.23}$$

The number of states of the "supersystem" for a given distribution $n_1, n_2, \ldots$ is given by:

$$\Omega_t(n) = \frac{(n_1 + n_2 + \ldots)!}{n_1! n_2! \cdot \ldots} = \frac{N_S!}{\prod_j n_j!} \tag{2.24}$$

The problem we initially set out to solve in studying the canonical system was to determine the probability of observing a quantum state E_j in a randomly chosen system from those comprising the canonical ensemble (or, in other words, the fraction of systems with state E_j).

Given a specific distribution $n_1, n_2, \ldots$, the probability of observing a quantum state E_j in a randomly chosen system from the ensemble is $\frac{n_j}{N_S}$, taking into account that there are many possible distributions given by N, V, N_S, and E_t.

The corresponding overall probability is an average of $\frac{n_i}{N_S}$ over all possible distributions, considering that each state of the supersystem is equiprobable. This implies that the weight assigned to each distribution is proportional to $\Omega_t(n)$ for each distribution.

Therefore, the probability for a given state E_j to be observed in any of the systems in the canonical ensemble is:

$$P_j = \frac{\overline{n}_j}{N_S} = \frac{1}{N_S} \cdot \frac{\sum_n \Omega_t(n) \cdot n_j(n)}{\sum_n \Omega_t(n)} \tag{2.25}$$

where $n_j(n)$ is the value of n_j in the distribution n. The averages of the energy and pressure over the entire system are:

$$\overline{E} = \sum_j P_j E_j \tag{2.26}$$

$$\overline{p} = \sum_j P_j p_j \tag{2.27}$$

where p_j is the pressure in state E_j, defined as: $p_j = -\left(\frac{\partial E}{\partial V}\right)_N$, with $-p_j dV = dE_j$ being the work done on the system in state E_j to produce a volume variation V.

An exact calculation of P_j is possible using a method called the Darwin-Fowler method, but it requires the use of very complex mathematical tools.

An alternative method, although not as rigorous, is applicable in the case of ensembles with a very large number of systems ($N_S \to \infty$) and is called the *method of the maximum term approximation*. In this case, an approximation is made based on the consideration that the most probable distribution is the one that dominates the averages of the system's properties. Let n^* denote the most probable distribution, corresponding to the highest degeneracy $\Omega_t(n^*)$: if N_S is large but finite, the distribution is of Gaussian type centered on $n = n^*$.

The maximum term approximation consists of replacing the entire summation referring to the average over the entire ensemble with the maximum term of the summation itself. In particular, the probability of the j-th distribution is:

$$P_j = \frac{\overline{n}_j}{N_S} \simeq \frac{1}{N_S} \cdot \frac{\Omega_t(n^*) n_j^*}{\Omega_t(n^*)} \tag{2.28}$$

where n_j^* is the maximum value in the distribution n_j.

By using this approach, the average value of n_j can be replaced with the value corresponding to the most probable distribution (which corresponds to the maximum value of Ω_t), and the problem becomes finding the maximum of the distribution, which can be calculated using the method of undetermined multipliers.

The most probable distribution provides the maximum value of Ω_t and, of course, of $\ln \Omega_t$;[7] to evaluate this, we use another approximation called Stirling's approximation, valid for very large numbers y:

$$\ln y! \simeq y \ln y - y \quad \text{if} \quad y \to \infty \tag{2.29}$$

Since $\Omega_t(n) = \frac{N_S!}{\prod_j n_j!}$, using Eq. 2.28, we find:

$$\ln \Omega_t(n) = \left(\sum_i n_i\right) \ln \left(\sum_i n_i\right) - \sum_i n_i \ln n_i \tag{2.30}$$

According to the method of undetermined multipliers, the set of n_j that correspond to the maximum value of $\ln \Omega_t(n)$ is found from the equations:

$$\frac{\partial}{\partial n_j}\left[\ln \Omega_t(n) - \alpha \sum_i n_i - \beta \sum n_i E_i\right] = 0 \tag{2.31}$$

where α and β are the undetermined multipliers.

For n_j^* corresponding to the most probable distribution, we find:

$$n_j^* = N_S \cdot e^{-\alpha} e^{-\beta E_j} \tag{2.32}$$

which depends on α and β.

Recalling Eqs. 2.22 and 2.23, we obtain:

$$N_S = \sum_j n_j^* = N_S \sum_j e^{-\alpha} e^{-\beta E_j} \tag{2.33}$$

$$e^{\alpha} = \sum_j e^{-\beta E_j} \tag{2.34}$$

Furthermore, since $E_t = N_S \overline{E}$, we have:

$$E_t = N_s \cdot \sum_j P_j E_j = \frac{\Omega_t(n^*) n_j^*}{\Omega_t(n^*)} E_j \tag{2.35}$$

[7] This is valid because $\ln x$ increases monotonically with x.

Equations 2.34 and 2.35 provide α and β as implicit functions of N and V (since $E_j = E_j(N, V)$).

Therefore, for the probability of the j-th state in the most probable distribution, we find:

$$P_j = \frac{n_j^*}{N_S} = \frac{e^{-\beta E_j(N,V)}}{\sum_j e^{-\beta E_j(N,V)}} \tag{2.36}$$

with β being positive. Consequently, from Eq. 2.36, it follows that the probability of observing a given quantum state decreases exponentially with increasing energy associated with it.

2.4.1.1 Thermodynamic Variables for the Canonical Ensemble

The thermodynamic quantities used to define a system can be either nonmechanical (temperature, entropy) or mechanical (pressure, energy E).

For a canonical system, using the first postulate, it is possible to replace the characteristic values of pressure P and energy E of the macroscopic ensemble with their corresponding average values $\overline{P}$ and $\overline{E}$.

Since the system is closed, considering the differential of E at constant N:

$$dE = \sum_j E_j dP_j + \sum_j P_j dE_j \tag{2.37}$$

where $dE_j = \left(\frac{\partial E_j}{\partial V}\right) NdV$ and $Pj = \frac{e^{-\beta E_j}}{\sum_j e^{-\beta E_j}}$; let's define

$$Q = \sum_j e^{-\beta E_j} \tag{2.38}$$

This term is called the *partition function of the canonical ensemble* or simply the *canonical partition function*. We obtain $\ln P_j = \ln\left(e^{-\beta E_j}\right) - \ln Q$. Equation 2.37 becomes:

$$dE = \frac{1}{\beta}\sum_j \left(\ln P_j + \ln Q\right) dP_j + \sum_j P_j \left(\frac{\partial E_j}{\partial V}\right)_N dV \tag{2.39}$$

Since $\sum_j P_j = 1$, we have $\sum_j dP_j = d\left(\sum_j P\right) = 0$; furthermore, $d\left(\sum_j P_j \ln P_j\right) = \sum_j P_j d\ln P_j + \sum_j \ln P_j dP_j = \sum_j \ln P_j dP_j$. Substituting these expressions into Eq. 2.39, we find:

$$dE = -\frac{1}{\beta} d\sum_j P_j \ln P_j + \sum_j P_j \left(\frac{\partial E_j}{\partial V}\right)_N dV \tag{2.40}$$

Using the first law of thermodynamics, we can write $TdS = dE + PdV$, and using Eq. 2.40, we can write:

$$TdS = -\frac{1}{\beta}d\left(\sum_j P_j \ln P_j\right) \tag{2.41}$$

$$P = -\sum_j P_j \left(\frac{\partial E_j}{\partial V}\right)_N \tag{2.42}$$

Using these relations, we find the expression for entropy:

$$S(N, V, T) = -k_B \sum_j P_j \ln P_j \tag{2.43}$$

Based on this equation, we find $\beta = \frac{1}{k_B T}$ through comparison;[8] from Eq. 2.43, we derive the expression for Helmholtz free energy, which represents the most appropriate thermodynamic potential for describing a closed system (with characteristic variables N, V, T):

$$A(N, V, T) = -k_B T \cdot \ln Q(N, V, T) \tag{2.44}$$

From the definition of dA:

$$dA = -SdT - PdV + \sum_\alpha \mu_\alpha dN_\alpha \tag{2.45}$$

we can derive:

$$S = -\left(\frac{\partial A}{\partial T}\right) V, N = kBT \cdot \left(\frac{\partial(\ln Q)}{\partial T}\right) V, N + kB \ln Q \tag{2.46}$$

$$P = -\left(\frac{\partial A}{\partial V}\right) T, N = kBT \cdot \left(\frac{\partial \ln Q}{\partial V}\right)_{T,N} \tag{2.47}$$

From Eqs. 2.43 and 2.36, we find the expression for entropy:

$$S(N, V, T) = k_B \ln Q + \frac{\overline{E}}{T} \tag{2.48}$$

[8] Through this definition, thermodynamic quantities regain their dependence on T, instead of E, which is more appropriate for closed and isothermal systems.

Finally, for the chemical potential of the i-th component, we have:

$$\mu_i = \left(\frac{\partial A}{\partial N_i}\right) T, V, N\alpha = i = -k_B T \cdot \left(\frac{\partial \ln Q}{\partial N_i}\right) T, V, N\alpha = i \tag{2.49}$$

and for the energy of the system:

$$E = ST + A = k_B T^2 \left(\frac{\partial \ln Q}{\partial T}\right)_{V,N} \tag{2.50}$$

For the canonical ensemble, therefore, the thermodynamic functions necessary to completely describe the thermodynamic system are available, in terms of microscopic properties, with the treatment just presented.

Another approach is based, rather than on deriving all possible microstates, on the energy level and the corresponding number of associated states. Let $\Omega_i(N, V)$ be the degeneracy of the i-th energy level.[9] For the partition function, we have:

$$Q(N, V, T) = \sum_{(states)} e^{-E_j(N,V,T)/k_B T} = \sum_{(levels)} \Omega_i(N, V) e^{-E_j(N,V)/k_B T} \tag{2.51}$$

The probability that the system exists in the energy level E is:

$$P_{(level,E)} = \Omega P_{\text{state}} = \frac{\Omega e^{-E/k_B T}}{Q} \tag{2.52}$$

Let's analyze Eq. 2.52 in two limiting cases:

1. For $T \to 0$, the lowest energy level E_1 is nondegenerate, so we have:

$$Q \to e^{-E_1/k_B T}\left(1 + \Omega_2 e^{-(E_2 - E_1)/k_B T} + \ldots\right) \to e^{-E_1/k_B T} \tag{2.53}$$

Thus, as $T \to 0$, the only possible state is the one with the minimum energy, so $S \to 0$, and the probability for the only nondegenerate state is:

$$P_j = \begin{cases} 1 & \text{if } j = 1 \\ 0 & \text{otherwise} \end{cases} \tag{2.54}$$

2. For $T \to \infty$, the relative effect of different E_j values on the Boltzmann factors is negligible, so $P_j(\text{state}) \to$ constant (independent of j). In other words, the

[9] In other words, in the list of all possible energy states $E_1, E_2, \ldots$, the same value E_i is found Ω_i times.

probability distribution over all states is uniform, leading to $S \to \infty$, assuming an infinite number of possible energy states.

2.4.2 Grand Canonical Ensemble

The grand canonical ensemble corresponds to an open and isothermal system, with fixed V, T, and μ. In other words, the thermodynamic system of interest is in thermodynamic equilibrium, with volume V, placed in a large thermal reservoir at a fixed temperature T, and it is open to exchange matter. In other words, the walls of the system allow the transport of matter and heat.

The reservoir is a tank in a thermostat at constant temperature T and ensures the different chemical species have chemical potentials $\mu_1, \mu_2, \ldots$. The composition of the system is given by the molar quantities of the n chemical species $N_1, N_2, \ldots$, which are values that fluctuate in time around their mean values $\overline{N}_1, \overline{N}_2, \ldots$.

Using a method similar to that used to derive the properties of the canonical ensemble, a supersystem is constructed by summing the constituent systems of the grand canonical ensemble. By replacing the time averages of the mechanical properties of the thermodynamic system with the ensemble averages of these properties, based on the first postulate, the probability factors (averaged over the ensemble) can be derived using the method of indeterminate multipliers. Finally, the values of these multipliers can be evaluated by comparing the thermodynamic and mechanical-statistical expressions for the mechanical variables.

In this case, unlike the canonical ensemble, the number N of molecules in the N_S systems of the ensemble is not fixed, but varies. For each different value of N, there is a different set of energy states $E_j(N, V)$. The quantum mechanical state of the suprasystem is fully determined when the value of N and the energy state $E_j(N, V)$ for each constituent system of the suprasystem are specified.

In a given state of the supersystem, let $n_j(N)$ be the number of systems containing N molecules in a particular energy state $E_j(N, V)$. The number of possible quantum states is (see Sect. 1.3):

$$\Omega_t(n) = \frac{\left[\sum_{j,N} n_j(N)\right]!}{\prod_{j,N} n_j(N)!} \tag{2.55}$$

Furthermore, conservation laws must be satisfied, which are entirely analogous to Eqs. 2.22 and 2.23.

Denoting:

$$\Xi = \sum_{j,N} e^{-\beta E_j(N,V)} e^{-\gamma N} \tag{2.56}$$

as the *grand canonical partition function*, we find, analogously to what was found for the canonical ensemble, $\beta = \frac{1}{k_B T}$ and $\gamma = \frac{\mu}{k_B T}$.

For the entropy S of the supersystem, i.e., for the grand canonical ensemble, we find:[10]

$$S = -k_B \sum_{j,N} P_j(N) \ln P_J(N) \tag{2.57}$$

For the probability of the j-th state, we have:[11]

$$P_j(N, T, V, \mu) = \frac{e^{-E_j(N,V)/k_B T} e^{-N\mu/k_B T}}{\Xi(V, T, \mu)} \tag{2.58}$$

where:

$$\Xi = \sum_{j,N} e^{-\beta E_j(N,V)} e^{-\mu N/k_B T} \tag{2.59}$$

is the complete definition of the grand canonical partition function. From this definition, it is possible to derive:

$$S = k_B T \left(\frac{\partial \ln \Xi}{\partial T} \right)_{V,\mu} + k_B \ln \Xi \tag{2.60}$$

$$N = k_B T \left(\frac{\partial \ln \Xi}{\partial \mu} \right)_{V,T} \tag{2.61}$$

$$p = k_B T \left(\frac{\partial \ln \Xi}{\partial V} \right)_{T,\mu} = k_B T \frac{\ln \Xi}{V} \tag{2.62}$$

As for the canonical ensemble, these definitions represent a complete set of relationships between macroscopic quantities (S, P, N, V,T) and the representative quantity of the microstates constituting the system, i.e., the grand canonical partition function.

2.4.3 Microcanonical Ensemble

The microcanonical ensemble is representative of an isolated thermodynamic system, with fixed E, V, and N. To derive the thermodynamic properties of the ensemble, characteristic properties of both the canonical and grand canonical ensembles are used. For example, the microcanonical ensemble can be seen as a

[10] The expression found for the entropy is entirely general and valid for every type of system.

[11] This expression is derived under the approximation of the maximum term.

degenerate canonical ensemble in which all systems have the same energy E. In other words, the only quantum states accessible to the system are those with energy E.

In this system, the fraction P_j of systems in a given quantum state (with energy E) is proportional to e^{-E/k_BT}; however, E is the same for all quantum states, which are in a number of $\Omega(N, V, E)$; therefore, P_j is the same for all Ω quantum states, so $P_j = 1/\Omega$. This leads to the expression:

$$S(N, V, E) = -k_B \sum_j P_j \ln P_j = k_B \ln \Omega(N, V, E) \tag{2.63}$$

This relationship, derived by Boltzmann, between the thermodynamic quantity S and the statistical-mechanical quantity Ω allows us to calculate other thermodynamic quantities of interest. Starting from the general form of the internal energy $E(S, V, N)$ in differential form:

$$dE = TdS - pdV + \sum_i \mu_i \cdot dN_i \tag{2.64}$$

Introducing the form 2.63:

$$dE = k_BT\, d\ln\Omega - pdV + \sum_i \mu_i dN_i \tag{2.65}$$

We can derive:

$$d\ln\Omega = \frac{dE}{k_BT} + \frac{p}{k_BT}dV - \sum_i \frac{\mu_i}{k_BT}dN_i \tag{2.66}$$

From which we immediately obtain:

$$\frac{1}{k_BT} = \left(\frac{\partial \ln\Omega}{\partial E}\right)_{V,N} \tag{2.67}$$

$$\frac{p}{k_BT} = \left(\frac{\partial \ln\Omega}{\partial V}\right)_{E,N} \tag{2.68}$$

$$-\frac{\mu_i}{k_BT} = \left(\frac{\partial \ln\Omega}{\partial N_i}\right)_{E,V,N_{\alpha\neq i}} \tag{2.69}$$

In classical thermodynamics, the functional relationships between the thermodynamic variables are independent of the environment (open system, closed system, isobaric, isothermal, etc.). In other words, the choice of independent thermodynamic variables is arbitrary and does not depend on the environment.

In statistical thermodynamics, the same conclusion is reached, as seen for entropy: regardless of the environment, we can choose any set and corresponding partition function to calculate thermodynamic properties, and the final result will be independent of the chosen set.

Equation 2.63 allows an interpretation of thermodynamic equilibrium and entropy in terms of internal disorder or order of the system. Considering that thermodynamic equilibrium for an isolated system corresponds to a condition of maximum entropy, from a microscopic perspective, the maximum entropy (and therefore thermodynamic equilibrium) corresponds to the condition with the maximum number of configurations, i.e., the condition of maximum disorder. This aspect will be discussed in more detail in the next paragraph.

2.4.4 Microscopical Interpretation of Entropy

The second law of thermodynamics, in the form expressed by Clausius ("The entropy of an isolated system tends to evolve toward a maximum value"), finds a brilliant interpretation through the atomic theory of matter, which states that bodies are composed of particles with translational, rotational, and vibrational motion.

From this perspective, the second law can be translated as follows: "Natural processes are processes of *mixing* of microscopic characteristics," meaning that a body tends to evolve in the direction where all properties of its elementary microscopic components tend to become uniform.

Some kinetic processes that have a final goal of thermodynamic equilibrium can be interpreted in terms of the natural tendency of mixing and achieving uniformity of microscopic properties over time. For example, this approach can be used to analyze the mixing of two fluids A and B in contact or the distribution of heat between two bodies at different temperatures in contact. In particular, in the latter case, what becomes uniform is not the microscopic concentration, i.e., the arrangement of molecules of different species in the microscopic lattice structure of the system, but the kinetic energy of the constituent particles, which becomes nearly uniform at equilibrium.

In reality, interpreting natural phenomena in terms of mixing emphasizes the significance of irreversibility, which represents the impossibility of spontaneously achieving the segregation of molecules of different species or energy states.

In general, we can distinguish two types of mixing processes: the dispersion of particles in space and the redistribution or dispersion of available energy among the particles themselves. The process of mixing is spontaneous (or natural) if the available energy is distributed over various quantum states in the "most disordered way possible."

In the mixing of two liquids A and B, the intermolecular interaction energy between binary molecular pairs (A-A, A-B, B-B) determines the equilibrium conditions. These conditions are achieved when, relative to the relative strengths of intermolecular forces between similar (A-A and B-B) and dissimilar (A-B) molecules, complete mixing would result in an increase in the potential energy of

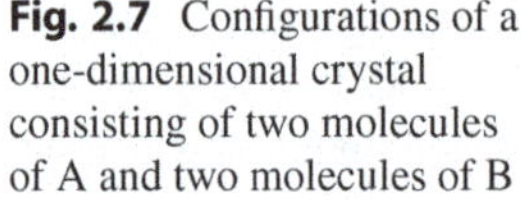

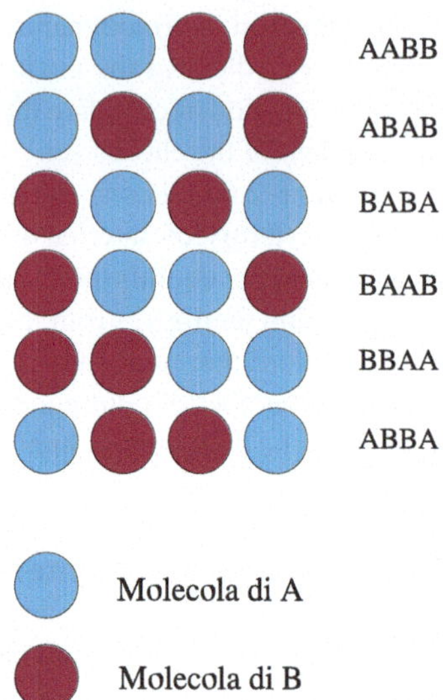

Fig. 2.7 Configurations of a one-dimensional crystal consisting of two molecules of A and two molecules of B

the system. Such energy would need to be subtracted from the kinetic energy of the molecules (translational, vibrational, and rotational motion) and would be less than the energy allocated to the distribution over the quantum levels associated with the motion state of the molecules. In this case, complete mixing would result in an increase in spatial or configurational disorder but a decrease in thermal disorder. The degree of miscibility at equilibrium depends on the concentrations of the two phases that correspond to the maximum overall disorder (configurational + thermal disorder).

For example, for a one-dimensional crystal, all possible configurations in which $N_A = 2$ molecules of A and $N_B = 2$ molecules of B can be placed correspond to all possible permutations of four objects, which is 4!; however, some configurations among the 4! possibilities are indistinguishable, particularly the 2! permutations of groups of identical molecules. The number of possible configurations will then be $\Omega = \frac{4!}{2!2!} = 6$ (see Fig. 2.7). In general, given $\bar{n} = \{n_1, n_2, \ldots\}$ indistinguishable objects (molecules), the number of configurations of such objects (distinguishable) is:

$$\Omega(\bar{n}) = \frac{(n_1 + n_2 + \ldots)!}{n_1!n_2!\cdots} \tag{2.70}$$

It is observed that the MACROSCOPIC state of maximum mixing corresponds to the largest number of configurations (MICROSTATES) possible, and therefore, it is also the most probable state, assuming the equiprobability of configurations (according to the second postulate).

Consequently, it can also be derived that natural processes (irreversible and spontaneous) always proceed in the direction that realizes the maximum number of possible configurations.

Further insight is needed regarding the Boltzmann expression for entropy 2.63; in it, entropy is simply referred to as the number of configurations available to the system, so it only has a *configurational* significance. However, thermal dispersion, i.e., the homogeneous distribution of kinetic energy, which corresponds to the *thermal* contribution of entropy, must also be taken into account. The balance between these two contributions must always result in a positive variational value for processes that occur spontaneously. Moreover, expression 2.63 establishes a strong connection between the entropy of a state and its corresponding probability. From this perspective, it can be stated that high-entropy states are also the most probable ones. For these reasons, the term Ω is also called *thermodynamic probability*.

Entropy is often associated with the concept of order/disorder, so an increase in entropy is associated with an increase in disorder. However, there are practical cases that seem to contradict this association, such as the spontaneous crystallization of an undercooled liquid in an adiabatic system.

This process involves a transition from a less ordered state (undercooled liquid) to a more ordered state (solid crystal). However, the overall entropy of the system increases because the kinetic (or thermal) energy increases, as the potential energy decreases due to the formation of the crystalline lattice. Therefore, the decrease in configurational entropy is compensated by the thermal contribution, resulting in a positive overall entropy change, as required for a spontaneous process.

In this sense, the following definition given by Guggenheim is very appropriate: "an increase in entropy corresponds to a 'dispersion' of the system over a higher number of accessible quantum states."

Entropy is also associated with the concept of *information*. In fact, spontaneous processes always proceed in the direction of increasing the number of accessible quantum states for a given system. This means that at equilibrium, the system is in a condition where it can assume a very large number of equally probable microscopic configurations. As a result, it becomes increasingly difficult to know in which configuration the system is, especially as it approaches the equilibrium condition. In this sense, it can be said that *natural processes evolve toward maximum entropy and minimum information available to the observer, in terms of the indeterminacy of the system's quantum state.*

Additional Resources and Recommended Literature

Good classical textbooks on statistical mechanics and thermodynamics are [6, 9]; a general textbook from principles on application is given in [**dill2010molecular**].

Proteins

3

A human being should be able to change a diaper, plan an invasion, butcher a hog, conn a ship, design a building, write a sonnet, balance accounts, build a wall, set a bone, comfort the dying, take orders, give orders, cooperate, act alone, solve equations, analyse a new problem, pitch manure, program a computer, cook a tasty meal, fight efficiently, die gallantly. Specialization is for insects.

R. Heinlein

Abstract

Why it is important to know this material?

The chapter entails the central knowledge about the protein structure and function, including the thermodynamic definition of protein stability and the kinetic mechanism of enzymatic activity on the basis of the corresponding molecular mechanisms.

What is the key idea?

The basic idea is that the protein structure and function depends on the protein stability and interactions with the environment.

What is necessary to know already?

It is important to have a clear definition of the protein stability in terms of Gibbs free energy and its decomposition in enthalpic and entropic terms, and their meaning. It is also necessary to have the basics of the reaction kinetics modeling, including the effect of catalysts.

L. Di Paola, *Fundamentals of Molecular Bioengineering*,
https://doi.org/10.1007/978-3-031-42022-1_3

3.1 Introductive Elements

In the following chapters, there will be frequent reference, concerning colloidal systems, to aqueous solutions (with a polar solvent) containing one or more solutes with high molecular weight, which will be indicated by the generic term "biopolymer."

The classes of substances in nature that have these characteristics are diverse (such as polysaccharides and nucleic acids, to name a few). However, in terms of variety, presence, and peculiar behavior, the compounds of greater interest here are proteins, which serve a wide range of functions. For this reason, they are also the most abundant biopolymers within the organism (constituting more than 70% of the dry cellular extract).

In particular, protein compounds play a key role in various physiological and pathological mechanisms. Proteomics, in particular, allows for the analysis of the entire production of proteins by an organism, with the aim of developing gene therapies that directly express the pharmacological protein of interest, produced according to the genetic instructions of the specific genome.

This discipline starts with identifying the gene pattern that determines the synthesis of a protein compound, which is then analyzed in terms of structure and biological function. Additionally, the identification of genetic variants of individual protein molecules, whether active or inactive from a biological standpoint, allows for highlighting the molecular aspects of many congenital or genetic diseases (e.g., amyloidosis, sickle cell anemia).

The function of protein compounds depends on the process of formation and the stability of the corresponding molecular structures. For this reason, modern computational techniques in structural biochemistry are increasingly focused on predicting and interpreting the three-dimensional conformations of these compounds (see Chap. 9).

In the next paragraphs, the general properties—chemical composition and molecular structure—of protein compounds will be described, emphasizing the close relationship between structure and the specific functions that these compounds are designed to perform.

3.2 The Amino Acids

The term "protein" derives from the Greek word $\pi\rho o\tau\varepsilon\iota o\sigma$, meaning primary. This term refers to the centrality of the function of such compounds within the organism.

Protein molecules are characterized by high complexity, both chemically and structurally. Their molecular weight varies from a few thousand Da (for proteins like insulin) to one million Da (for myosin). In general, compounds with a molecular weight lower than 4000 Da are referred to as *peptides*.

From a strictly chemical point of view, proteins are copolymers of monomers called *amino acids*. These constituent elements have a typical structure shown in

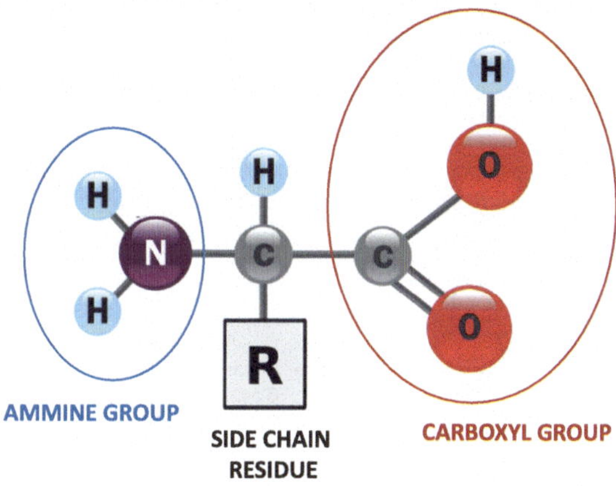

Fig. 3.1 Typical molecular structure of amino acids

Fig. 3.1. As you can see, they are characterized by an α carbon[1] to which an amino group and a carboxyl group are attached, giving the molecule a dual nature of acids (carboxyl group $-COOH$) and bases (amino group $-NH_2$). They differ from each other by the $-R$ group, called the residue , which determines their name and also their hydrophobic or hydrophilic nature.

Figure 3.2 shows the 20 amino acids commonly found in nature, divided into polar, acidic, basic, and hydrophobic categories. The large number of possible monomers that make up peptide polymers is the basis for their corresponding wide variety, which makes them the most versatile and widespread class of biopolymers in living organisms.

Amino acids combine with each other through a condensation reaction, resulting in a peptide bond (Fig. 3.3). This bond is highly stable and has a high binding energy because the $-R$ groups contribute to its protection, providing a steric energy barrier to the breakage of the bond. The peptide bonds that hold the amino acids together in polypeptide and protein compounds represent the first and most important structural level of the compound. They can be cleaved by specific enzymes, usually involved in digestion or bacterial attack processes.

3.3 Protein Structure

The understanding of the biological activities of protein compounds involves an analysis of the three-dimensional structure that these molecules assume in physiological environments.

[1] The asymmetric structure of the α carbon makes amino acids optically active molecules (it is interesting to note that amino acids found in nature are almost exclusively levorotatory).

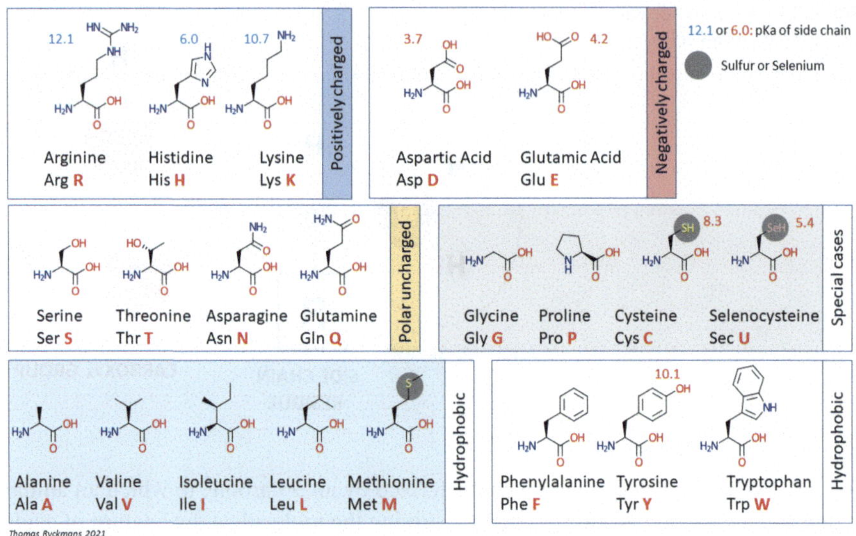

Fig. 3.2 Amino acids divided by chemical characteristics: hydrophobic amino acids are shown in cyan, polar amino acids in yellow, acidic amino acids in red, and basic amino acids in blue. Special cases are in gray. (reprinted with permission from https://commons.wikimedia.org/wiki/File:Proteinogenic_Amino_Acid_Table.png)

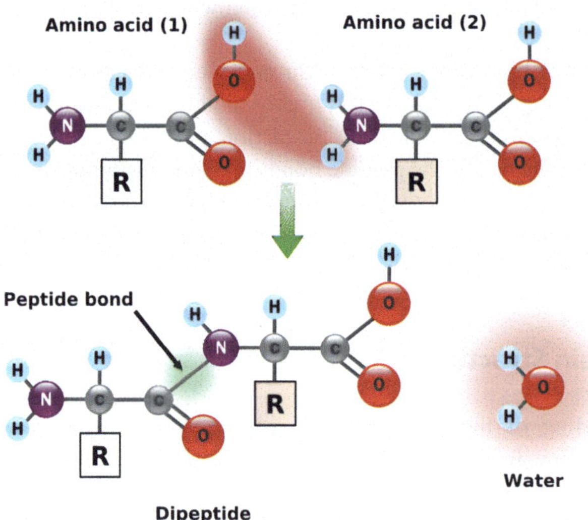

Fig. 3.3 Formation of the peptide bond by the reaction of condensation of two amino acids (reprinted with permission from https://chem.libretexts.org/Courses/University_of_Kentucky/UK%3A_CHE_103_-_Chemistry_for_Allied_Health_%28Soult%29/Chapters/Chapter_13%3A_Amino_Acids_and_Proteins/13.2%3A_Peptides)

Proteins, as they are expressed in vivo, have a well-defined structure called the *native* structure. The conformation of a protein molecule is determined by the mutual attraction or repulsion of the residues present within the molecule.

Protein structures are organized hierarchically into different levels: primary, secondary, tertiary, and quaternary structures. These levels also correspond to the various states in the dynamics of forming the final structure in the complex process called *folding*, which transforms the molecule from a disordered state, as directly expressed by ribosomes, to a three-dimensional shape corresponding to maximum biological activity.

3.3.1 Primary Structure

The *primary structure* of a protein is defined as the sequence of amino acids in the polypeptide chain; by convention, the sequence is counted from the amino end.

The experimental determination of the sequence is carried out by chemical or enzymatic degradation of the peptide chain, followed by analysis of the resulting amino acids obtained from the lysate. This structural level is determined by the covalent peptide bonds and already contains all the necessary information for the formation of the final three-dimensional structure.

The primary structure represents the first, simplest level of protein structure, but it is also the most robust. Only highly aggressive processes such as enzymatic digestion or exposure to very high temperatures can break the residue backbone of peptide bonds that form the primary chain of amino acids.

3.3.2 Secondary Structure

The secondary structure of a protein describes the local structural units that constitute the nuclei around which the three-dimensional structure is built. These units, in the process of folding, result from interactions between neighboring residues in the chain.

There are two main classes of secondary structure that are found in most proteins: α-helices and β-sheets (Fig. 3.4). The α-helix structure (Fig. 3.4, right) is mainly held together by hydrogen bonds between adjacent or distant $N - H$ and $C = 0$ groups in the peptide chain.

Hydrogen bonds are also involved in the formation of β-sheets (Fig. 3.4, left), which only form in regions where the amino acid glycine is abundant.

Secondary structures have the following characteristics:

- They are determined by the primary structure.
- They describe only local regions of the peptide chain.
- They are stabilized by a large number of hydrogen bonds. Although each individual hydrogen bond is a weak interaction, their abundance can strongly stabilize a structure (cooperative binding).

Fig. 3.4 Secondary structures in proteins: β-sheets (left) and α-helices (right) (reprinted with permission from https://en.wikipedia.org/wiki/Protein_secondary_structure)

- Since α-helices and β-sheets are stabilized by hydrogen bonds, any factor that interferes with the formation of these bonds can disrupt the corresponding secondary structures, e.g., factors such as heat (thermal agitation causes fluctuations in the various groups that counteract hydrogen bonding), pH changes, and high ionic strength (high concentrations of H^+ ions and other ions alter the polarity of the environment, interfering with hydrogen bond formation).

3.3.3 Tertiary and Quaternary Structure

The tertiary structure of a protein refers to its conformation in space when immersed in a specific environment. This structural level owes its stability to a combination of intramolecular interactions, primarily involving amino acid residues:

- Hydrogen bonds between polar residues.
- Ionic bonds between charged residues.
- Hydrophobic interactions between nonpolar groups.
- Covalent bonds: Cysteine residues contain a sulfhydryl group (-SH) capable of forming a covalent bond, known as a disulfide bond, with analogous residues in other regions of the polypeptide chain in the presence of oxidizing agents. This disulfide bond stabilizes the three-dimensional structure of the protein (Fig. 3.5).

  The tertiary structure can be altered by various factors that interfere with the intramolecular interactions that hold the protein structure together, including changes in temperature, pH, and ionic strength.

The loss of protein tertiary structure is referred to as protein denaturation, which means the loss of the protein's native structure, which is the functionally active three-dimensional conformation. Denaturation can be reversible or irreversible,

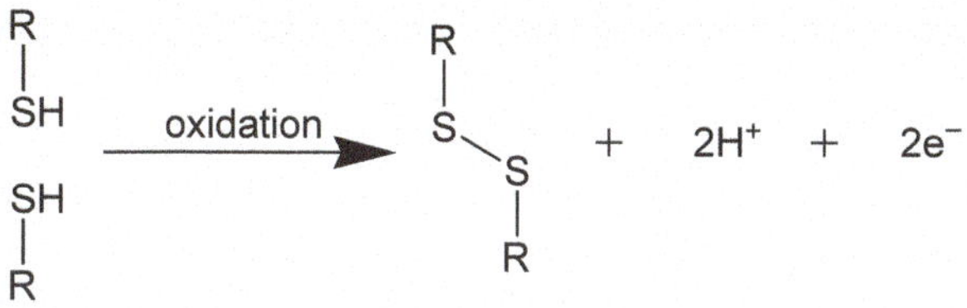

$$R\text{-}SH + HS\text{-}R \xrightarrow{\text{oxidation}} R\text{-}S\text{-}S\text{-}R + 2H^+ + 2e^-$$

Fig. 3.5 Disulfide bond (reprinted with permission from https://commons.wikimedia.org/wiki/File:Disulfide-bond.png)

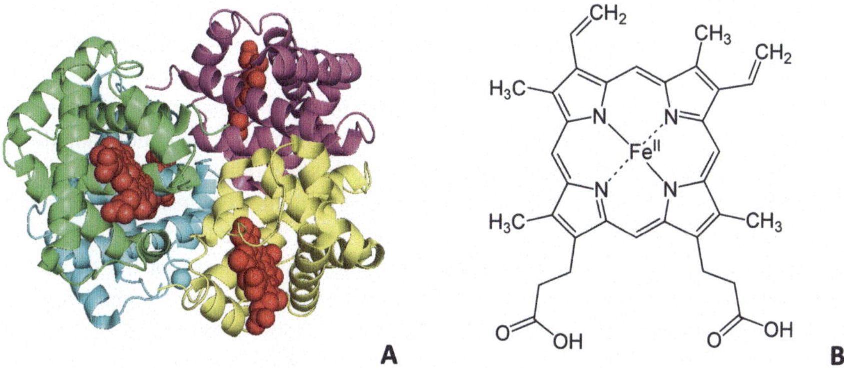

Fig. 3.6 Three-dimensional structure of hemoglobin: (**a**) the α chains are in green and magenta, the β in blue and yellow, and in red spheres the heme groups; (**b**) the structural formula of the heme group, in charge for the binding of hemoglobin with the respiratory gases

depending on whether the structure returns to its initial form after the denaturing factor is removed.

Based on their molecular shape, proteins can be classified as follows:

- *Fibrous proteins*: They have an elongated shape and are poorly soluble in water and aqueous solutions (e.g., collagen, keratin, silk fibroin).
- *Globular proteins*: They have a spherical shape, are much more soluble in water than fibrous proteins, and generally contain a high proportion of α-helices (e.g., lysozyme, bovine serum albumin).
- *Proteins with an intermediate structure* (e.g., fibrinogen): They have intermediate shape and physicochemical properties between fibrous and globular proteins.

The quaternary structure of a protein describes the interactions between different peptide chains that constitute the protein. Some proteins, such as hemoglobin (Fig. 3.6), are composed of multiple amino acid chains, each having its own three-dimensional structure, and they are held together by the same types of interactions that stabilize the tertiary structure of a single peptide chain.

3.4 Protein Function and Relationship with Structure

The function and activity of a protein depend on its protein structure, which, as mentioned earlier, relies on the amino acid composition, i.e., the primary structure. In fact, even a variation in the sequence of only a few amino acids can lead to dramatic changes in the final three-dimensional structure. This complex allows the substrate molecule to be transformed into the product, which exhibits low activity with the enzyme's active site and is subsequently released.

The function and activity of a protein depend on its protein structure, which, as mentioned earlier, is determined by the amino acid composition, i.e., the primary structure. In fact, even a variation in the sequence of just a few amino acids can lead to dramatic changes in the final three-dimensional structure. This was postulated by the American biochemist Anfinsen in 1961 in his thermodynamic hypothesis regarding the relationship between amino acid composition and the final three-dimensional structure of proteins. According to his hypothesis, all the necessary information for the biomolecule to reach its native structure is stored in the primary structure.

This position implies the uniqueness of a stable conformation for the biomolecular compound, which is assumed independently of other factors. Supporting this theory is the observation that for many proteins, a single amino acid variation in the primary sequence produces a dramatic variation in the three-dimensional conformation. A well-known example is sickle cell anemia, where a single amino acid differs in the hemoglobin protein compared to physiologically active hemoglobin. However, the final molecular structure is profoundly different in the pathological form, leading to instability in physiological conditions and the formation of harmful aggregates that deform red blood cells, giving them the characteristic sickle shape.

Proteins perform functions within living organisms, and among these functions, catalysis is of fundamental importance. Protein compounds that act as biocatalysts are called enzymes. They differ from inorganic catalysts in several ways:

- Enzymes are highly specific, meaning that each enzyme catalyzes one and only one reaction.
- Enzyme levels can be regulated according to the organism's needs.
- Enzymes can be inhibited by various factors, including those that cause denaturation of the protein structure. Once inhibition has occurred, the enzyme still acts as a specific catalyst for the original reaction but with a much slower reaction rate.
- Various factors control enzymatic activity, related to specific groups (notably sugars and lipid groups) that dramatically modify the protein structure and, therefore, the enzymatic activity.

The high specificity and efficiency of enzymes can be explained by the way they interact with the reacting molecules, commonly referred to as the *substrate*. One of

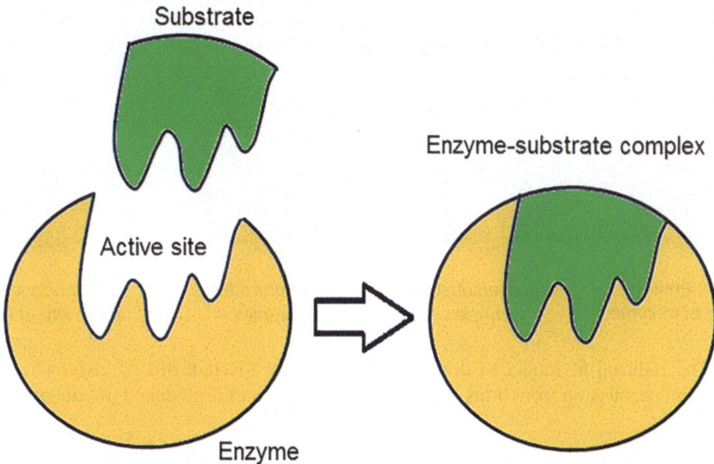

Fig. 3.7 Lock-and-key model to describe the molecular mechanisms of enzymatic catalysis (reprinted with permission from https://commons.wikimedia.org/wiki/File:Lock_and_key.png)

the earliest models proposed to explain the high efficacy of enzymatic catalysis is the *lock-and-key model* (Fig. 3.7).

According to this model, each enzyme molecule has few active sites on its surface. These sites are concave regions of the protein surface that exhibit significant structural and chemical affinity for substrate molecules, quantifiable in terms of an attractive potential that leads to the formation of a *substrate-enzyme complex*.

A more realistic model is called the *induced-fit* model, indicating a greater elasticity of the surface region that acts as the active site, which can explain the aforementioned effects. Similar to a glove, the active site can adapt to small changes in shape and size (see Fig. 3.8. This also explains how slight pH variations, leading to small deformations of the active site's structure, do not significantly alter enzymatic activity. As previously mentioned, the wide diversity in composition and structure of protein compounds underlies their broad spectrum of functions. Therefore, the following classification is rather limited compared to the complete range of functions expressed by these biopolymers:

- *Enzymes*—biological catalysts: Most reactions in biological systems require nonaggressive conditions (near-neutral pH and low temperatures). These conditions are provided by catalysts with high specificity, which are protein molecules with active sites on their molecular surface. The presence of these catalysts typically increases the reaction rate by at least a million-fold.
- *Storage*: Various ions, small molecules, and other metabolites are stored by forming complexes with protein molecules. For example, oxygen is stored in muscles through myoglobin, while iron in the liver is bound to ferritin.
- *Transport*: Proteins are involved in the transport of particles ranging in size from electrons to macromolecules. Transport function is often closely related

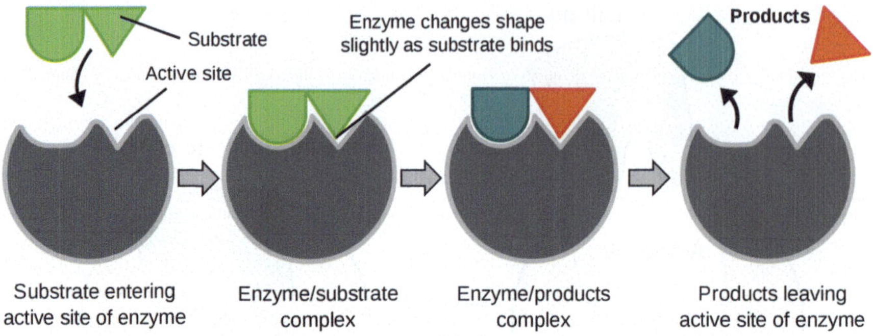

Fig. 3.8 The induced-fit model to describe the molecular mechanisms of enzymatic catalysis (reprinted with permission from https://it.m.wikipedia.org/wiki/File:Induced_fit_diagram.svg)

to storage function. For example, transferrin transports iron to the liver, where it is stored by forming complexes with ferritin, as mentioned earlier. Similarly, hemoglobin is responsible for transporting oxygen from the pulmonary alveoli to tissues, where it is stored through complex formation with myoglobin. A specific transport function is performed by membrane proteins, which control the transport of various substances to and from the cell.

- *Messenger molecules*: Proteins play a key role in transmitting various signals (electrical or chemical) to enable communication between distant regions. They are involved in the transmission of nerve impulses by acting as receptors for different low molecular weight substances that bridge with nerve molecules. Hormonal activity is also dependent on proteins, enabling the coordination of organ activity. Hormones and their corresponding receptors are predominantly protein compounds.
- *Antibodies*: The immune system relies on the production of antibodies, which are proteins that specifically bind to foreign agents such as bacteria and viruses. Regulation: The information required for protein synthesis is stored in genes, which are DNA sequences containing instructions for synthesizing a particular amino acid sequence. The precise regulation of gene expression is controlled by other protein substances, ensuring the appropriate levels of protein compounds.
- *Structural proteins*: Muscles, the skeleton, and other biological tissues, such as hair or nails, largely consist of proteins. Collagen, for example, is one of the main components of connective tissue and is also the most abundant protein in vertebrates (approximately a quarter of the protein content in mammalian organisms is represented by collagen).

3.5 Protein Folding and Denaturation

Protein denaturation (or unfolding) is a process in which the native structure of proteins undergoes a substantial change without breaking the primary (peptide) covalent bonds that form the residue backbone. In other words, denaturation involves changes in the secondary and tertiary structure but not the primary structure.

Protein denaturation is a phenomenon that occurs in our daily lives (it is the basis for cooking protein-rich foods) and plays an important role in technology. Protein compounds are used as enzymes or represent the final product in biotechnological and pharmaceutical industrial processes.

According to a simple scheme, denaturation can be considered as the reverse process of protein folding , according to the two-state transition model described by the following equilibrium reaction:

$$N \underset{}{\overset{K_D}{\rightleftharpoons}} D \tag{3.1}$$

Here, N represents the native form and D represents the denatured form. This simplification is applicable to moderately sized proteins that follow simple folding/unfolding pathways.

According to this model, there is a perfect reciprocity between the folding and denaturation processes. This representation allows studying the thermodynamics of folding by inducing the denaturation of the protein structure of interest. In other words, to determine the stability and resistance of a protein molecular structure, it is necessary to break it and measure the energy required for the rupture.

To thermodynamically analyze the denaturation process according to the kinetic scheme 3.1, it is necessary to define the reference state and the properties of the forms involved. The denatured form D, which also represents the reference state, is assimilated to the so-called random coil. This ideal form is not represented by a single structure but rather a class of configurations (a statistical ensemble of configurations) where the constitutive elements (residues) are randomly arranged in space, although still bonded together.

Once this reference state is determined, it is possible to define the molecule's stability in terms of the change in Gibbs free energy, $\Delta G_N^0 = G_N^0 - G_D^0$, which is associated with the folding process (*folding Gibbs free energy*). If the value of ΔG_N^0 is negative and high in absolute value, the corresponding native structure N is stable, whereas å if the value of ΔG_N^0 is positive, the corresponding structure is unstable. This situation may arise when environmental conditions do not permit the existence of the protein molecule in its native structure.

The native protein structure compound is only slightly more stable than the denatured forms at atmospheric pressure and room temperature. The characteristic values are $\Delta G_N^0 = G_N^0 - G_D^0 = -RT \cdot \ln K_N = -10 \div -5 \frac{kcal}{mole}$ for small protein compounds, corresponding to equilibrium constants $K_N = 10^4 \div 10^7$.

This data indicates that natural proteins have evolved by achieving a compromise between stability and adaptability. In fact, a high structural stability ($\Delta G_N^0 \ll 0$) would correspond to a high molecular rigidity, rendering the protein molecule unfit to adapt to environmental conditions and, therefore, biologically inactive. In fact, many synthetic proteins exhibit high stability but are biologically inactive.

The prediction of the three-dimensional protein structure from the amino acid sequence is based on the calculation of configurational free energy , which has $3N$ degrees of freedom, where N is the number of atoms in the structure. This complex potential function has many local minima, which represent stable three-dimensional configurations. Among these configurations, the one assumed by the molecule in its physiological environment corresponds to the maximum biological activity, ensuring that the surface active sites adopt the conformation most suitable for their function. The difference in free energy between the value corresponding to the minimum and that relative to the random coil is ΔG_N^0, and it measures, as previously mentioned, the stability of the structure corresponding to the energy minimum.

The native three-dimensional structure of a protein is the result of the tendency of hydrophobic residues, which are poorly compatible with the polar physiological environment, to avoid contact with water molecules. Consequently, the native structure can be represented as a two-layer sphere: an internal hydrophobic core surrounded by a hydrophilic shell. This phenomenon of segregating hydrophobic residues in the internal region of the molecule to avoid contact with the polar aqueous solvent is called the hydrophobic (or solvophobic) effect .

To this representation, the presence of hydrophobic regions on the hydrophilic surface should be added. These regions "tolerate" contact with water due to the layer of solvent molecules covering the molecule, thanks to the favorable interaction ($\Delta H \ll 0$) with the superficial hydrophilic residues, which are also close to the surface hydrophobic regions. These hydrophobic regions located on the surface of protein molecules often represent the added value of the protein compounds themselves, as they frequently constitute the active binding sites with other molecules. On the other hand, hydrophilic residues often serve in a less specific manner only to increase the solubility of the protein in its native form in aqueous solution.

At this point, it is appropriate to define the hydrophobicity of a generic compound quantitatively as the variation in Gibbs free energy ΔG_{id} related to the transfer of the compound from a polar environment (typically, water) to a nonpolar environment, or vice versa. For example, Table 3.1 shows *Tanford's hydrophobicity scale* , based on the measurement of amino acid solubility in polar (water and ethanol) and nonpolar (dioxane) solutions, as well as the evaluation of the corresponding free energy of transfer from the polar phase to the nonpolar phase. Negative values for nonpolar amino acids correspond to their higher solubility in hydrophobic organic solvent solutions, while positive values for polar, basic, and acidic amino acids indicate a decrease in solubility during the transition from the hydrophilic to the hydrophobic environment.

Table 3.1 Hydrophobicity of amino acids: Tanford's scale

Amino acid	hydrophobicity (Kcal/mol)
Arg	3.0
Asp	2.5
Glu	2.5
Asn	0.2
Lys	3.0
Gln	0.2
His	−0.5
Ser	0.3
Thr	−0.4
Tyr	−2.3
Gly	0
Pro	−1.4
Cys	−1.0
Ala	−0.5
Trp	−3.4
Met	−1.3
Phe	−2.5
Val	−1.5
Ile	−1.8
Leu	−1.8

The process of protein denaturation can proceed either reversibly or irreversibly, depending on the characteristics of the environment in which the protein chain is immersed, i.e., the protein structures modified after the application of a denaturing agent can restore their native conformation if the agent is removed. In this case, the process is reversible; otherwise, it is irreversible, and the corresponding structures are irreversibly compromised both in terms of structure and biological activity.

In general, for each denaturing agent , it is possible to identify a "intensity" limit of that agent[2] below which the process can be reversed by removing the agent itself. Beyond this limit, on the other hand, the protein structures are irreversibly compromised: in fact, following the unfolding of the chains, hydrophobic residues belonging to different chains interact, promoting the formation of stable aggregates.[3]

[2] The intensity of denaturing agents coincides with the concentration for chemical agents, temperature, or electromagnetic field intensity for thermal or radiative denaturation.

[3] For a long time, it was wondered why hydrophobic residues did not give rise, during the physiological folding phase, to the formation of similar aggregates: in reality, during the expression of protein compounds, one of the functions of support molecules, such as *chaperonins* , is precisely to "protect" hydrophobic residues from unwanted attractive interactions, such as intermolecular ones, thus favoring the formation of intramolecular bonds.

Referring to scheme 3.1, the denaturation equilibrium constant is defined as:

$$K_D = \frac{[D]}{[N]} = \frac{1}{K_N} \tag{3.2}$$

Indicating by $[P]$ the total protein concentration, applying a simple balance for the protein species yields:

$$[P] = [N] + [D] \tag{3.3}$$

The fractions of molecules in denatured and native forms are therefore $f_N = \frac{[N]}{[P]}$ and $f_D = \frac{[D]}{[P]}$, respectively. At this point, it is possible to rewrite the denaturation equilibrium constant in terms of fractions as:

$$K_N = \frac{f_N}{f_D} = \frac{1}{K_D} \tag{3.4}$$

Knowing that $f_N = 1 - f_D$, it is easy to derive the dependence of the fraction of protein in the denatured form on the denaturation equilibrium constant:

$$K_D = \frac{f_D}{1 - f_D} \qquad \rightarrow \qquad f_D = \frac{K_U}{1 + K_U} \tag{3.5}$$

The main characteristics of the denatured state are that it is extremely flexible, and therefore, its residues are widely exposed to the solvent, as well as the peptide backbone. Similar to the native state, the denatured macrostate is composed of a distribution of microstates, and this distribution is much broader than that in the native state. Furthermore, all possible spatial configurations (microstates) have approximately the same energy, making them equally probable.

The denaturation constant K_D can be related to the standard Gibbs free energy change associated with the denaturation process ΔG_D^0[4] through the van't Hoff equation :

$$\Delta G_D^0 = G_D^0 - G_N^0 = -RT \log K_D \tag{3.6}$$

The Gibbs free energy can, in turn, be written in terms of enthalpic and entropic contributions:

$$\Delta G_D^0(T) = \Delta H_D^0(T) - T \cdot \Delta S_D^0(T) \tag{3.7}$$

[4] The standard conditions correspond, in this case, to the temperature T of interest, atmospheric pressure, and the relative concentrations in the reference system (in this case, the Henry reference, which corresponds to the infinite dilution of the component of interest $x_i^0 \rightarrow 0$).

The transformation is also associated with a change in specific heat, which generally depends on temperature:

$$\Delta c_{PD}(T) = c_{PD}(T) - c_{PN}(T) \tag{3.8}$$

The dependence of enthalpy and entropy on temperature is evaluated from known values of the two potentials at a reference temperature T_0:

$$\Delta H_D^0(T) = \Delta H_D^0(T_0) + \int_{T_0}^{T} \Delta c_{PD}\, dT \tag{3.9}$$

$$\Delta S_D^0(T) = \Delta S_D^0(T_0) + \int_{T_0}^{T} \frac{\Delta c_{PD}}{T}\, dT \tag{3.10}$$

If the variation in specific heat with temperature can be neglected (e.g., when considering a relatively narrow temperature range), Eqs. 3.9 and 3.10 become, respectively:

$$\Delta H_D^0(T) = \Delta H_D^0(T_0) + \Delta c_{PD} \cdot (T - T_0) \tag{3.11}$$

$$\Delta S_D^0(T) = \Delta S_D^0(T_0) + \Delta c_{PD} \cdot \ln\left(\frac{T}{T_0}\right) \tag{3.12}$$

If it is assumed that Δc_{PD} remains constant with temperature.[5]

In addition (Eq. 3.11), the term Δc_{PD} represents the positive slope of the line describing the dependence of the denaturation enthalpy change $\Delta H_D(T)$ on temperature. $\Delta H_D(T)$ increases with increasing temperature (endothermic process), and this increase is more pronounced with higher values of Δc_{PD}.

In general, the thermodynamic potentials related to the transformations of protein molecules in an aqueous environment can be separated into two terms: one related to the actual structural changes and the other related to the reorganization of water molecules around the biomolecules. Regarding this latter contribution, it has been found that the corresponding term for Gibbs free energy ΔG_w is practically zero, meaning there is a perfect compensation between the entropic and enthalpic terms. This is because a reorganization of water molecules occurs around the hydrophobic regions exposed to the solvent, resulting in a loss of their degrees of freedom and a negative entropic contribution ΔS_w, which corresponds to a positive contribution to the Gibbs free energy. In turn, the enthalpic term ΔH_w is negative because, upon the formation of the solvation layer, the hydrated biomolecular structure establishes

[5] This assumption is reasonable for the denaturation of a large number of protein structures, especially within the narrow temperature range around the environment where the major conformation changes underlying thermal denaturation occur.

more favorable interactions with the surrounding solvent compared to the absence of a solvation layer.

The denaturation process is irreversible if stable aggregates of denatured molecules form due to the interaction of exposed hydrophobic residues after the native structure is disrupted. Assuming a two-state denaturation hypothesis, this process can be represented by the following kinetic scheme:

$$N \underset{k_i}{\overset{k_d}{\rightleftharpoons}} D \overset{k_a}{\rightarrow} A \tag{3.13}$$

where A represents the aggregate of denatured molecules.

Using this reference scheme, we can evaluate how the concentrations of the three species present, N, D, and A, vary, for example, in a closed reactive system where n_0 moles of native molecules N are loaded. The material balances for the three species are:

$$\begin{cases} \frac{dc_N}{dt} = -k_d \cdot c_N + k_i \cdot c_D \\ \frac{dc_D}{dt} = k_d \cdot c_N - (k_i + k_a) \cdot c_D \\ \frac{dc_A}{dt} = k_a \cdot c_D \end{cases} \tag{3.14}$$

These three linear homogeneous differential equations have a closed-form solution with the initial conditions $c_N(0) = c_{N0} = \frac{n_0}{V}$, $c_D(0) = 0$, and $c_A(0) = 0$. In particular, the general solution of the system composed of the first two equations, in vector form, is:

$$\dot{\mathbf{c}} = A \cdot \mathbf{c} \rightarrow \mathbf{c} = \mathbf{c}(0) \cdot \exp(M \cdot t) \tag{3.15}$$

where M is the dynamic matrix representing the linear system 3.14. Figure 3.9 shows the time evolution of the concentrations of the protein species, which can be obtained from the solution 3.15 of the system 3.14.

3.5.1 Denaturing Agents

As explained in the preceding paragraphs, the native structure of proteins is the most stable in the polar physiological aqueous environment (water + salts). It is stabilized by non-covalent bonds that depend on the characteristics of the interacting residues and the environment in which they interact. Consequently, any factor that modifies this environment can cause denaturation.

On the other hand, the bonds that are formed have to "counteract" the energy of thermal agitation, which gives the amino acid groups random motion (translation, rotation, and vibration) that opposes the formation of intramolecular bonds. For this reason, one of the major denaturing agents is an increase in temperature, which dramatically changes the three-dimensional structure of the protein molecule.

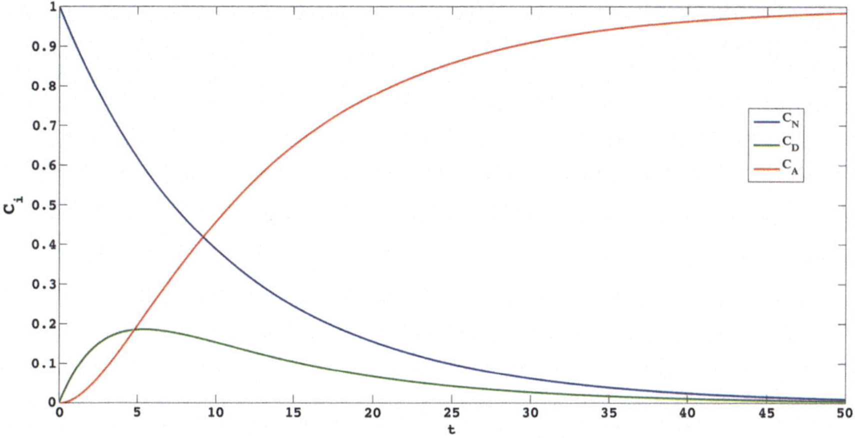

Fig. 3.9 Concentration of different protein forms in the irreversible denaturation process of Eq. 3.13

Moreover, the increased average velocity of denatured molecules also leads to a higher frequency of collisions between denatured molecules, resulting in irreversible aggregation. Notably, thermal denaturation is the basis of food cooking, as it makes the protein molecules in the food product more digestible and deactivates bacterial activity by denaturing bacterial proteins.

From a thermodynamic perspective, denaturation can be reversible or irreversible, which essentially depends on the tendency of denatured molecules to form stable aggregates or not.

Denaturation by *alcohols* is based on altering the polarity of the environment in which the protein molecule is immersed. In particular, this alteration affects hydrogen bonds, which have the same nature as the bonds between water and alcohol molecules. Alcoholic denaturation is the basis of the effectiveness of alcohol-based disinfectants.

Changes in pH and ionic strength also lead to conformation variations in protein structure. These changes primarily affect ionic bonds (salt bridges) between electrostatically charged residues. The same effect occurs in the presence of heavy metal ions, which are often used in topical disinfectants but can also cause intoxication when ingested or in contact with the body.

Furthermore, heavy ions and reducing agents act on the disruption of disulfide bonds, which are the only class of covalent bonds that stabilize the three-dimensional structure of proteins.

Strong protein denaturants include nonalcoholic organic compounds (benzenic compounds, linear hydrocarbons, aldehydes, ketones, etc.) that are highly hydrophobic. The presence of such molecules creates a strongly nonpolar environment, which radically changes the nature of intramolecular interactions in protein molecules.

As seen previously, the native three-dimensional structure of a protein results from the tendency of hydrophobic residues, which are less compatible with the polar physiological environment, to avoid contact with solvent molecules. Consequently, the native structure can be represented as a two-layer sphere: an inner, highly non-polar spherical core surrounded by a hydrophilic spherical shell. This simplification overlooks an important aspect: in nature, the primary structures of proteins usually have a large majority of hydrophobic residues. When the polarity of the solvent decreases abruptly, hydrophobic interactions become significantly weaker, leading to chain unfolding and the formation of a denatured structure (random coil).

The denaturation mechanism differs for urea and guanidine chloride. These molecules directly interact with the peptide bond, altering the overall energy of the protein structure. This causes a reorganization of the structure and, consequently, a loss of the original native integrity.

In the case of denaturation by chemical agents, an empirical linear relationship is observed between the Gibbs free energy of denaturation ΔG_D and the concentration of the denaturing agent $[d]$:

$$\Delta G_D([d]) = \alpha - \beta \cdot [d] \tag{3.16}$$

where $\alpha = \Delta G_D([d = 0])$ and β are two experimentally determined constants.

3.5.2 Cooperativity and Folding

The complexity of biological processes stems from the fact that they are based on the formation or breaking of a large number of (covalent and non-covalent) bonds. These processes are said to be *cooperative* if the overall energy is not simply the sum of the energies of individual bonds. If the overall energy is lower than the sum of the individual contributions, it is called *positive cooperativity*, meaning the phenomenon is energetically less costly and therefore more likely to occur. Conversely, if the required energy is higher, it is called *negative cooperativity*.

Many biological processes exhibit this behavior, including the formation of three-dimensional conformations (protein folding), the protonation of polyelectrolytes (each charge released from the polyelectrolyte experiences increasing difficulty detaching from the molecule), the solvation of biomolecules, and the formation of protein-ligand complexes (which will be extensively discussed later).

Specifically, cooperative bonds are involved in protein folding. From an energetic standpoint, if the formation of a certain type of bond corresponds to the Gibbs free energy ΔG, n bonds of the same type generally yield a contribution different from $n \cdot \Delta G$. Protein folding typically exhibits positive cooperativity because when two spatially distant regions approach each other to form the first intramolecular bond, the energy required for this operation also includes the energetic contribution to bringing all the residues in the two sections closer. Therefore, subsequent bonds formed between residues from the two different sections somehow benefit from the

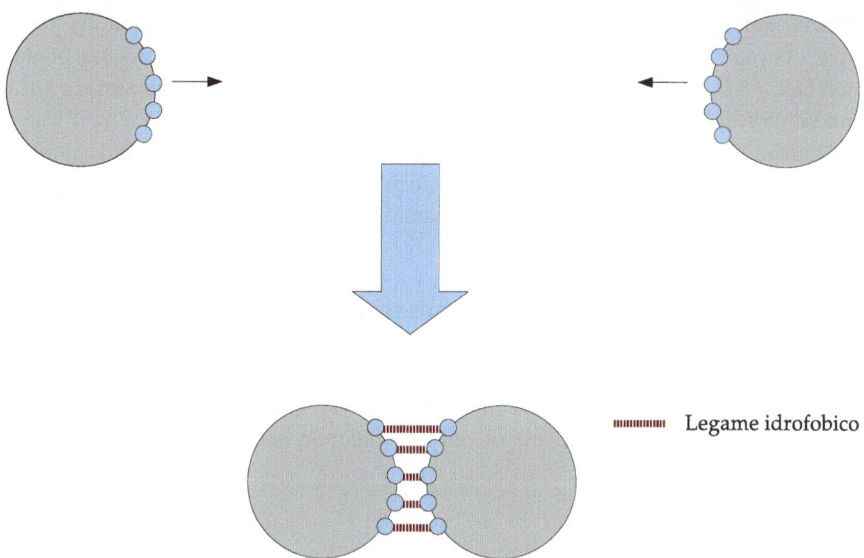

Fig. 3.10 Formation of cooperative hydrophobic bonds

work done during the approach for the formation of the first bond, and they will be energetically less costly than the first bond.

The cooperativity of folding is particularly evident in the two-state folding scheme (Eq. 3.14); as soon as the first bond is formed, all the others form in cascade, leading to the formation of the native structure in a single step. On the other hand, regarding the reverse process of denaturation, as soon as the first bond (typically hydrophobic in nature) stabilizing the native globular structure is broken, all the other bonds break in cascade, and the completely unfolded form is formed in a single step starting from the native structure.

Similarly, the nature of intermolecular cooperative bonding can be illustrated as in Fig. 3.10.

The two molecules have sites on their surfaces that can interact with each other, forming bonds (e.g., hydrophobic residues in an aqueous environment). For this to happen, the molecules must approach each other at a certain distance, and the energy required for the formation of the first bond includes the energy spent on bringing the two molecules closer. The bonds established subsequently leverage the fact that the molecules are already in the proper position. Therefore, the resulting stability will be greater than that resulting from the first bond since the energetic cost for the formation of subsequent bonds after the first one will be lower.

Finally, another process that exhibits this cooperative behavior is the solvation of a macromolecule in an aqueous solution, particularly if it contains hydrophobic regions on its surface. A hydrophobic solute, in fact, repels water molecules and does not solvate in an aqueous solution (positive Gibbs free energy of solvation $\Delta G_{solv} > 0$) . On the contrary, a molecule with an electrolytic or polar nature attracts water molecules once it is placed in a solution. The formation of a layer of water molecules (solvation layer) promotes the solvation process of the molecule ($\Delta G_{solv} < 0$). However, the hydrophobic regions on the surface of a protein molecule are surrounded by extensive hydrophilic regions that attract water molecules. These water molecules, in turn, promote the solvation of the hydrophobic regions. This is a clear cooperative phenomenon, resulting from the heterogeneity and chemical complexity of protein compounds.

Additional Resources and Recommended Literature

The basic reference of biochemistry is given in [14]; the biophysical characterization of protein molecular structure and function is provided in [3, 4, 13]. The characterization of molecular mechanisms of enzyme function and activity is available in [2].

Intermolecular Forces and Potentials

4

A friend should bear his friend's infirmities.
Julius Caesar, Act 4, Scene 3

W. Shakespeare

Abstract

Why it is important to know this material?

The intermolecular forces causally determine the properties of matter. This chapter highlight the general models for intermolecular forces and how this interaction potentials can be leveraged to derive the property of gases and liquids.

The description of the commonest intermolecular potentials is provided in the chapter.

What is the key idea?

The intermolecular forces and corresponding potentials can used to derive macroscopic properties of fluid phases, such as the virial coefficients for gas and osmotic pressures.

What is necessary to know already?

The general concepts of statistical thermodynamics are required to connect intermolecular potentials and macroscopic properties.

4.1 Introduction

The interactions between molecules, involving both attractive and repulsive forces, govern the behavior of a system comprising these molecules.

Configurational properties of a system are those properties depending on the interaction forces between the molecules rather than on the characteristics of the isolated molecules. For example, the latent heat of vaporization of a liquid is a

L. Di Paola, *Fundamentals of Molecular Bioengineering*,
https://doi.org/10.1007/978-3-031-42022-1_4

configurational property (the energy required to break the non-covalent bonds that hold the molecules together in the liquid phase), while the specific heat of a low-pressure gas is definitevely not configurational (the gas can be considered ideal, with noninteracting molecules, so the specific heat only depends on the thermal agitation of the individual molecules).

If the resultant of the interactions between two molecules has spherical symmetry and depends only on the distance between the centers of mass of the two molecules r (as in the case of Coulomb's law), the intermolecular potential $W(r)$ for conservative forces can be derived from the well-known relationship:[1]

$$F = -\frac{dW}{dr} \tag{4.1}$$

which, in integral form, becomes:

$$\int_r^{\infty} F\,dr = -\int_r^{\infty} dW = -W(\infty) + W(r) = W(r) \tag{4.2}$$

In the previous expression, it is assumed that the potential becomes negligible as $r \to \infty$, meaning that the interactions between molecules are zero when the molecules are infinitely far apart. In other words, $W(r)$ represents the work required to move two molecules from a center-to-center distance r to an infinite distance apart.

By fixing a radial coordinate r with the center of the spherical molecule as the origin, the attractive forces are directed opposite to this coordinate and are therefore negative, while the repulsive forces are positive. According to the definition given by Eq. 4.1, the work done by the attractive forces (opposite to the fixed coordinate) is negative, while the work done by the repulsive forces is positive (Fig. 4.1).

The medium through which the interaction potential operates strongly modifies the nature and intensity of the interactions. In the case of gas molecules, intermolecular interaction between different molecules occurs in a vacuum, while solutes in a liquid solution are subject to intermolecular interaction forces acting in a solvent

[1] A spherically symmetric potential can only be applied to molecules with spheroidal shape and homogeneous surface properties (e.g., charges or hydrophobicity). In the case of more complex geometries or surfaces with nonuniform properties, additional spatial variables are necessary for a realistic description.

In the case of a non-spherically symmetric potential, the force vector $\mathbf{F}$ acting on a point in space, whose position is defined by the vector $\mathbf{x}$, can be derived from the intermolecular potential $W(\mathbf{x})$ for conservative force fields as:

$$\mathbf{F} = -\nabla W(\mathbf{x}).$$

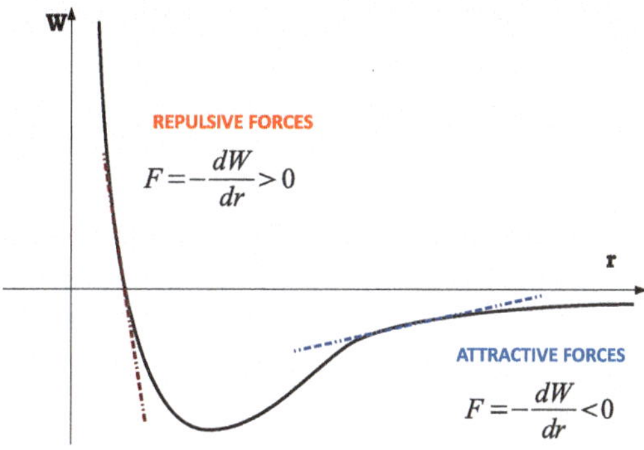

Fig. 4.1 Molecular interaction potential

medium.[2] Thus, while in the first case an absolute definition of interaction forces is made, in the second case, the forces are mediated by the presence of the solvent.

A well-known and immediate example is provided by electrostatic forces. In a vacuum, two point charges Q_1 and Q_2 at a distance r interact with each other through the Coulomb force, directed along the line joining the two charges. This force follows the well-known Coulomb's law :

$$f_0 = \frac{Q_1 Q_2}{4\pi\varepsilon_0} \cdot \frac{1}{r^2} \tag{4.3}$$

where ε_0 is the permittivity in a vacuum, equal to $8.85 \cdot 10^{-12}$ F/m. If the same charges interact in a dielectric medium, Eq. 4.3 becomes:

$$f = \frac{Q_1 Q_2}{4\pi\varepsilon} \cdot \frac{1}{r^2} \tag{4.4}$$

where ε is the permittivity in the dielectric medium, given by $\varepsilon = \varepsilon_0 \cdot \varepsilon_r$, with the contribution of the dielectric medium represented by the average relative permittivity ε_r, which reflects the average dielectric properties of the medium.

In general, the presence of a solvent modifies the interactions between two molecules compared to interactions in a vacuum. It is even possible for two molecules to attract each other in a vacuum while repelling each other in a solvent medium. While in a vacuum, the two molecules approach each other due to attractive forces that depend solely on the nature of the molecules, in the solvent medium, the work required to push away solvent molecules in the approach path must be subtracted from the work provided by these forces. This mechanism is similar to the

[2] In this case, interactions between solvent molecules and between solvent molecules and macromolecules can be neglected due to the large difference in size between the two types of molecules. This assumption neglects the molecular nature of the solvent, considering it as an isotropic and homogeneous medium, similar to a "heavy vacuum." This assumption is known as the *mean-field* approximation.

advancement of vacancies during diffusion in solids with defects and vacancies. If the latter term is larger than the former, the net interactions become repulsive.

Another phenomenon to consider when introducing a solute into a solvent is the rearrangement of solvent molecules around the solute molecules, thereby modifying the inherent structure of the solvent (solvation) . As a result, solute molecules in the solvent generally have different physical characteristics, such as charge and dipole moment, compared to when they are in a vacuum.

In general, the process of solubilizing a molecule can be schematized as a process consisting of two stages:

1. The creation of a cavity in the solvent with suitable dimensions and shape to accommodate the solute molecule. This contribution is predominantly entropic and depends on the shape and size of the molecule to be accommodated in the cavity, regardless of the chemical nature of the molecule itself.
2. The formation of bonds between the solute molecule and the solvent molecules that are exposed on the surface of the cavity. This contribution is predominantly enthalpic and depends on the chemical nature of the solute and solvent, as well as the number of solvent molecules involved, i.e., the size of the cavity.

4.2 Definition of Intermolecular Potentials

In the case of extremely dilute systems (such as binary mixtures with a very dilute solute or gases at very low pressure), the spatial distance between molecules is so large compared to the size of the molecules themselves that they can be considered point-like and the intermolecular interactions can be neglected. This molecular model corresponds, in different systems, to the well-known laws of perfect gases and van't Hoff for osmotic systems , which respectively describe the volumetric behavior of gases at very low densities and the osmotic pressure of dilute liquid solutions.

However, in systems with higher molecular densities, the intermolecular potential cannot be neglected. In this case, the conditions of ideality are not satisfied, and the aforementioned laws are no longer applicable.

The first assumption that is no longer valid when considering the non-ideality of the system is that molecules are no longer considered point-like, but with a finite diameter (in the case of spherically symmetric molecules), neglecting other contributions to the intermolecular potential. In this case, molecules can be approximated as billiard balls (spherical symmetry) composed of an infinitely rigid material, meaning that the intermolecular interactions are zero for distances r greater than the diameter σ of the spheres (rigid sphere diameter), while for $r \leq \sigma$, the repulsive force is infinite. This potential, called the *hard-sphere intermolecular potential* , is formally expressed as:

$$W(r) = \begin{cases} \infty & r \leq \sigma \\ 0 & r > \sigma \end{cases} \tag{4.5}$$

If attractive forces (dispersion forces) are also present, the expression of the potential that is obtained (*Sutherland potential*) is:

$$W(r) = \begin{cases} \infty & r \leq \sigma \\ -\varepsilon \left(\frac{\sigma}{r}\right)^6 & r > \sigma \end{cases} \tag{4.6}$$

The parameters of the Sutherland potential model are ε, the depth of the potential well, the rigid sphere diameter σ (also known as the collision diameter), and the interaction parameter K.

More generally, intermolecular forces can be classified as attractive (dispersion, specific, hydrogen bonding, osmotic attraction) and repulsive (rigid sphere, electrostatic). In general, the overall potential can be defined when both attractive and repulsive contributions are present:

$$W(r) = W_{repulsive}(r) + W_{attractive}(r) = \frac{A}{r^n} - \frac{B}{r^m} \tag{4.7}$$

where A, B, n, and m are positive constants. This function exhibits a minimum at the distance value $r_{\min}$ with a depth of ε. Equation 4.7 can be rewritten as:

$$\frac{W(r)}{\varepsilon} = \frac{(n/m)^{1/(n-m)}}{n-m} \cdot \left[\left(\frac{\sigma}{r}\right)^n - \left(\frac{\sigma}{r}\right)^m\right] \tag{4.8}$$

where $\varepsilon = -W(r_{\min})$ and σ is the intermolecular distance such that $W(\sigma) = 0$.

Figure 4.2 shows *Mie's intermolecular potential* ($m = 6$) for different values of n.

For $n = 12$ and $m = 6$, Eq. 4.8 becomes the *Lennard-Jones intermolecular potential* (Fig. 4.3):

$$\frac{W(r)}{\varepsilon} = 4\left[\left(\frac{\sigma}{r}\right)^{12} - \left(\frac{\sigma}{r}\right)^6\right] \tag{4.9}$$

The two parameters required to define this potential are an energy-related parameter (ε) and a geometric parameter (σ). The repulsive branch, which is very steep, occurs for $r \leq \sigma$, while the attractive branch approaches zero as $r \to \infty$.

Although the repulsive branch is steep, it still has a finite slope, which means that if two molecules collide with sufficiently high energy, they can deform, and the two centers of mass can approach distances less than σ. All repulsive potentials that have this characteristic are called *soft-sphere intermolecular potentials*.

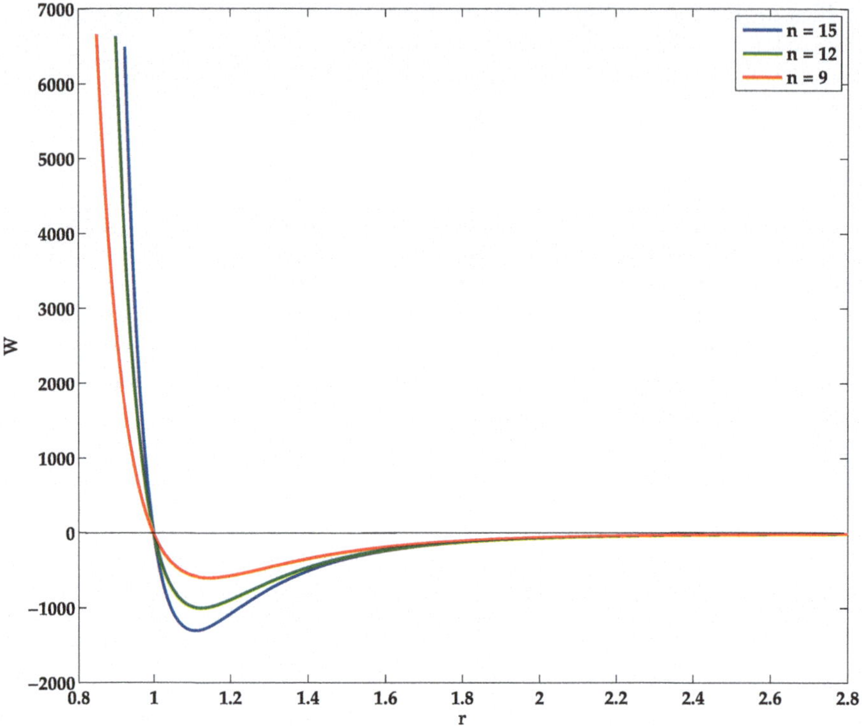

Fig. 4.2 Mie's intermolecular potential

A brutally simplified form of the Lennard-Jones potential, which is less accurate but more commonly used, is the *square-well intermolecular potential*::

$$W(r) = \begin{cases} \infty & r \leq \sigma \\ -\varepsilon & \sigma < r \leq k\sigma \\ 0 & r > k\sigma \end{cases} \tag{4.10}$$

This potential has three adjustable parameters: σ, k, and R, which allow for an adequate description of many molecular systems whose interaction potential includes both repulsive and attractive terms.

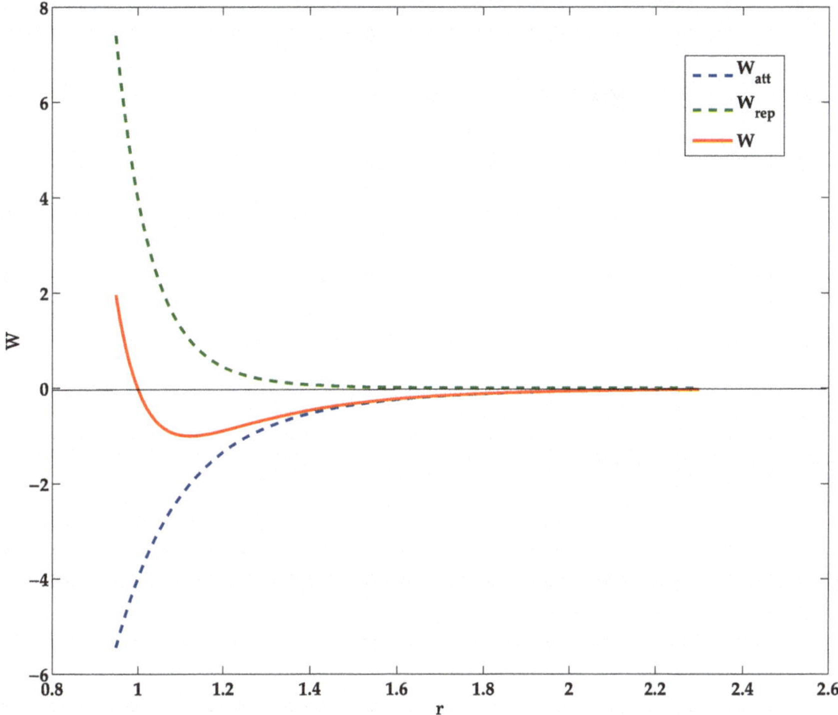

Fig. 4.3 Lennard-Jones intermolecular potential

4.3 Structure and Properties of Liquids

The description of the molecular structure of liquids is a rather recent field in the physics of condensed matter, which relies on the latest and most complex experimental techniques for defining the molecular properties of this state of matter, which possesses intermediate properties between those of gases and solids.

The definition of the liquid state starts from the critical density ρ_c, which is known for every pure compound: a liquid has a density $\rho \geq \rho_c$. Furthermore, defining the *compressibility* as $\chi = \left(\frac{\partial \rho}{\partial(\beta P)}\right)_\beta$, where $\beta = \frac{1}{k_B T}$, a liquid is such that $\chi \leq \frac{1}{50}$.

The internal structure of a liquid can be measured experimentally using X-ray diffraction and scattering of electromagnetic radiation (light or X-rays) or neutron beams. The quantity that defines the structure of a liquid is called the *radial distribution function*, and it is defined using the following procedure.

One of the molecules of the liquid is taken as a reference (see Fig. 4.4, panel a); starting from the center of this molecule, a radial coordinate r is considered; an infinitesimal volume $d\tau$ is considered at a distance r from the center of the molecule,

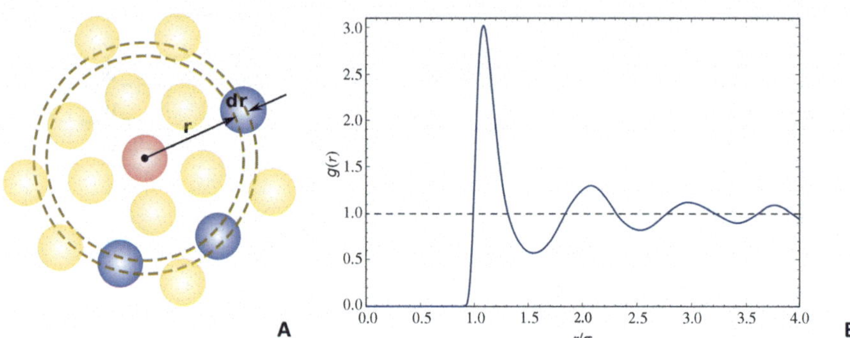

Fig. 4.4 Radial distribution in a liquid: (**a**) Molecular structure of a liquid; (**b**) radial distribution for a Lennard-Jones fluid (reprinted with permission from https://en.wikipedia.org/wiki/Radial_distribution_function and https://en.wikipedia.org/wiki/File:Lennard-Jones_Radial_Distribution_Function.svg)

and the number of molecules contained in this volume is not simply $\rho d\tau$ (where ρ is the average number density), but a different number that depends on how the molecules are arranged in space.

The radial distribution function $g(r)$ is defined in such a way that the number of molecules present in the spherical shell of thickness $d\tau$ at a distance r from the center of mass of the molecule is $g(r)\rho d\tau$; in other words, $g(r)$ represents the factor by which the average density ρ is multiplied to provide the local density at r.

This function has some interesting properties:

- $g \rightarrow 0$ as $r \rightarrow 0$; in fact, at small distances, all molecules are impenetrable.
- $g \rightarrow 1$ as $r \rightarrow \infty$; at large distances, the observed local density is equal to the average density (no longer influenced by the molecule taken as the reference).

Figure 4.4, panel b, shows a comparison between the typical radial distribution function for a Lennard-Jones liquid.

Additional Resources and Recommended Literature

A very general perspective on the intermolecular forces and how they relate to the matter properties is given in [11]. A good description of the molecular structure of liquids is given in [6].

Volumetric Properties of Fluids: Equations of State

5

I dwell in Possibility — A fairer House than Prose — More numerous of Windows — Superior — for Doors

E. Dickinson

Abstract

Why it is important to know this material?

The equation of the state of gases and solutions is the basic determinant of the behavior of these systems. The parallelism between gas pressure and osmotic pressure in liquid solutions highlights a unitary approach that enables the description of gas mixtures and liquid phase solutions at the macroscopic and microscopic level in similar ways.

What is the key idea?

The equations of state for gases and liquid solutions provide an operational framework to define the thermodynamic properties of these systems and their phase equilibria.

What is necessary to know already?

Classical thermodynamics tools are essential to utilize state equations in the thermodynamic analysis of fluid systems.

5.1 Introduction

Classical thermodynamics is an engineering science in its deep roots, stemmed out of the need to quantitatively describe the relationship between the properties of an object or a region of space (system). The great challenge that led to the formulation of the principles of thermodynamics was to extend a similar logic to the laws and formalism applied to the study of deformations in solids to the fluid phases,

L. Di Paola, *Fundamentals of Molecular Bioengineering*,
https://doi.org/10.1007/978-3-031-42022-1_5

which, at the beginning of the nineteenth century, lacked a rigorous and systematic description.

The historical need arose at the beginning of the Industrial Era when it became necessary to consolidate empirical knowledge related to phase transitions and energy transmission during compression-expansion cycles into a body of quantitative and general laws. This allowed for the design of devices that could exploit these properties, moving beyond the artisanal practice of constructing machinery capable of empirically performing similar functions.

In the two centuries that followed, thermodynamics has reached high levels of abstraction, but the general laws formulated within that framework and context have remained valid. Molecular thermodynamics provides tools to relate these laws to the microscopic properties of the system.

Today, we face various needs, just as we did two centuries ago during an energy crisis, and we seek to solve them using new materials with properties defined at the nanoscale. We aim to design biocatalysts (enzymes) capable of transforming biological waste (biomass) into biofuels. This drives us to find optimized systems for these new production methods, and we need to know in detail the molecular nature of the systems to accurately predict their behavior. Once again, we rely on general laws, where the terms are defined based on molecular parameters. We have simply increased the molecular detail of laws that are universally valid.

In this context, it is essential to define the volumetric properties of systems, that is, to understand how a system behaves when pressure or density changes at constant temperature. When defining volumetric properties, we typically distinguish colligative properties, which depend only on the number of molecules and not on their chemical or physical nature. In this classification, there is implicitly a molecular model: colligative properties are properties of strongly diluted systems, where the intermolecular interaction potential can be neglected.

The characteristics of fluid systems are surprisingly similar, in terms of molecular models, when passing from compressible gas systems to nearly incompressible liquid systems. For both types of systems, it is possible to define a pressure, which represents mechanical pressure and can alter the density of gas systems, as well as osmotic pressure in liquid systems when considering mixtures composed of a solvent with one or more solutes. The molecular models used to describe the relationships between pressure (osmotic pressure), concentration, and temperature are similar and lead to the same type of functional relationships. In this chapter, these relationships (equations of state) and their application and interpretation in fluid systems will be discussed.

5.2 Equations of State for Ideal and Nonideal Systems

A thermodynamic system is defined by several macroscopic variables (temperature, volume, number of moles, pressure, osmotic pressure). Equations of state are functional relationships between these quantities. These equations are used to describe the thermal properties of the system (how the system reacts to changes in temperature) and the volumetric properties of the system, which are derived for isothermal systems and expressed in terms of the relationship between pressure and density at a fixed temperature.

Equations of state for gas systems are functions that relate the number of moles of compounds in the system n_i, pressure P, temperature T, and volume V:

$$F(n_i, P, T, V) = 0 \tag{5.1}$$

For liquid solutions, the same type of relationships exists between the number of moles n_i of the i-th solute, the corresponding osmotic pressure π_i (which will be defined later in the chapter), and temperature:

$$F(n_i, P, T, V) = 0 \tag{5.2}$$

Equations of state can be classified as *empirical* or *theoretical*. Theoretical equations can be derived from molecular models that describe the interactions between the molecules constituting the system. These models allow for generalization to different systems. For example, the simplest molecular model, which considers the system as composed of noninteracting point-like molecules, leads to the same expression for the equation of state of ideal gases and to the van't Hoff equation for describing osmotic pressure as a function of solute concentration. Although gas molecules interact in a vacuum, while solute molecules interact in a medium, the solvent (mean-field interaction), the result is analogous. As shown later, this analogy holds for other molecular models and their corresponding equations of state (Fig. 5.1).

In other words, gas systems and (dilute) liquid solutions can be described using similar tools because the molecular laws that translate into macroscopic descriptive

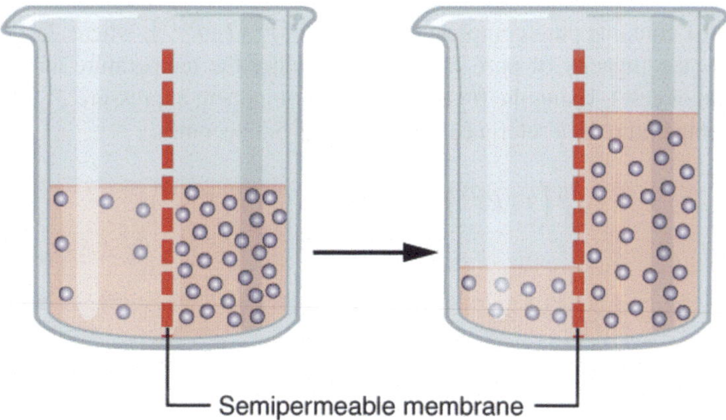

Fig. 5.1 Osmotic equilibrium (reprinted with permission from https://en.wikipedia.org/wiki/Osmosis)

quantities of the system state are analogous. Gas molecules, whether interacting or not, generate pressure on the container surfaces in which they are contained. Similarly, solute molecules that interact in an environment composed of solvent molecules can be represented by continuous quantities (mean-field description).

5.3 Osmotic Pressure of Colloidal Systems

As discussed in Sect. 2.1, in gases the pressure P is the result of molecular collisions with the walls of a container held at temperature T and volume V, containing N molecules. The interactions between gas molecules, which occur in a vacuum, completely define the molecular-level system and can be translated into an equation of state for the gas under the given conditions.

Similarly, in solutions containing solutes, there is an associated pressure called osmotic pressure, which, similar to gas systems, depends on the intermolecular interaction field between solute molecules in the solution.

For osmotic pressure, an operational definition can be given based on the description of the so-called osmotic equilibrium. Consider a system at constant temperature T consisting of two cells separated by a semipermeable membrane[1] A practical case that can be analyzed with this model is the dialysis of a solution containing macromolecules, using a membrane with a lower cutoff size than the macromolecules. The membrane allows only the passage of component 1 (solvent) but not component 2 (solute), which is confined to only one of the cells (cell II), while cell I contains only component 1.[2]

Thermodynamic equilibrium (osmotic equilibrium) for component 1 across the membrane is expressed by the equality of solute fugacities in the two cells:

$$f_1^{(I)} = f_1^{(II)} \tag{5.3}$$

Since cell I contains pure component 1, $f_1^{(I)} = f_1^0\left(T, P^{(I)}\right)$, where $f_1^0\left(T, P^{(I)}\right)$ represents the fugacity of pure component 1 under the temperature and pressure conditions of cell I. Using the fugacity of pure component at pressure $P^{(II)}$ and the system temperature as a reference for cell II, Eq. 5.3 becomes:

$$f_1^0\left(T, P^{(I)}\right) = x_1^{(II)} \gamma_1^{(II)} f_1^0\left(T, P^{(II)}\right) \tag{5.4}$$

[1] We assume the membrane to be rigid so that a pressure difference ΔP can be established between the two cells.

[2] One practical case that can be analyzed with this model is the dialysis of a solution containing macromolecules, using a membrane with a lower cutoff size than the macromolecules.

The relationship between the fugacities of pure component 1 in the two cells (at different pressures $P^{(I)}$ and $P^{(II)}$) is given by the Poynting expression:[3]

$$f_1^0\left(T, P^{II}\right) = f_1^0\left(T, P^{I}\right) \cdot \exp\left[\frac{\widehat{v}_1^L\left(P^{II} - P^{I}\right)}{RT}\right] \tag{5.5}$$

where $\hat{v}_1^L$ is the molar specific volume of component 1 in the liquid phase at the assigned temperature. Applying Eqs. 5.5–5.4, we find:

$$1 = x_1^{II}\gamma_1^{II} \exp\left[\frac{\widehat{v}_1^L\left(P^{II} - P^{I}\right)}{RT}\right] \tag{5.6}$$

and thus:

$$P^{(II)} - P^{(I)} = \Delta P = \pi = -\frac{RT}{\widehat{v}_1^L} \cdot \ln\left(x_1^{II}\gamma_1^{II}\right) \tag{5.7}$$

The pressure difference between the two cells π is called the *osmotic pressure*, and it is the pressure difference that must be established for thermodynamic equilibrium (equality of chemical potentials, which translates into the equality of fugacities) between the two cells through the membrane.

From a molecular point of view, similar to what was described for gas pressure, osmotic pressure is the result of the collisions of solute molecules (component 2) with the membrane walls in cell II; this results in a macroscopic pressure difference between the two sides of the membrane (this term is absent in cell I, where only solvent 1 is present).

5.4 Virial Coefficients and Intermolecular Potentials: MacMillan-Mayer Theory

The virial equation of state provides an explicit relationship between the parameters of the equation (virial coefficients) and the intermolecular interaction potential. The general n-th virial coefficient can be obtained as an integral expression (in space) of the potential between n-particle clusters (interaction involving n bodies).

The McMillan-Mayer theory has been developed to relate the non-ideality of a gas to the corresponding intermolecular potentials based on the virial expansion of the compressibility factor z (in terms of the gas's hydrostatic pressure). The objective is to define a molecular theory for interpreting the virial coefficients in

[3] The pressure in cell II will be greater than the pressure in cell I, where the pure component is present, to balance the fugacity term in cell I, which contains pure solvent, with an activity $a^{(II)} = \gamma^{(II)} \cdot x^{(II)}$ less than unity in cell II.

terms of molecular potentials. This theory is extended to describe dilute colloidal systems, in order to utilize the results obtained for the series expansion of the hydrostatic pressure of gases and the osmotic pressure of colloidal solutions.

This approach is based on a theorem known as the virial theorem , which describes the overall characteristics of interacting particle systems in terms of time averages.

Series Expansion for a Dilute Gas

For an extremely dilute gas, the well-known ideal gas law holds: $p/(\rho k_B T) = 1$, where $\rho = V/N_{\text{molecules}}$ is the number density of the gas. At higher densities (corresponding to higher pressures), the equation of state can be corrected as follows:

$$\frac{p}{\rho k_B T} = 1 + B_2(T)\rho + B_3(T)\rho^2 + \ldots \tag{5.8}$$

This relationship, as discussed earlier, is known as the virial expansion.[4] $B_2(T)$, $B_3(T)$, ..., are called the second, third, ... virial coefficients, respectively, and they correspond to two-, three-, ... body interactions. These coefficients vary from gas to gas and depend on the intermolecular interactions. It should be noted that these coefficients depend only on the macroscopic variable T.

In simple terms, when the gas is sufficiently dense, the gas molecules spend more time in close proximity to each other. This interaction becomes important first at the pairwise level ($\rightarrow$ represented by $B_2(T)$), then at the triplet level ($\rightarrow$ represented by $B_3(T)$), and so on, as the density increases. In other words, as the density ρ increases, interactions involving a larger number of molecules become increasingly important, and higher-order terms are included in the series expansion. The series converges for $T < T_C$ and $\rho \leq \rho_G$, where ρ_G is the density of the saturated vapor at the given temperature.

The n-th virial coefficient $B_n(T)$ is defined as:

$$B_n(T) = \frac{1}{(n-1)!}\left(\frac{\partial^{n-1}\frac{p}{\rho k_B T}}{\partial \rho^{n-1}}\right)_{T,\rho=0} \tag{5.9}$$

The dependence of the second virial coefficient on the intermolecular potential is derived based on a statistical model that uses an appropriate statistical ensemble to separate the terms of the thermodynamic functions related to two-, three-, four-, ... body interactions. The grand canonical ensemble (discussed in Sect. 2.1) is the statistical ensemble that allows for this separation. In fact, the grand canonical

[4] Not all real dilute gases can be studied using a virial expansion; a notable case is that of ionic plasmas.

partition function can be expressed as a series expansion in terms of the *absolute activity* $\lambda = e^{\mu/(k_BT)}$ (see Sect. 2.3):

$$\Xi(V, T, \mu) = Q(0, T, V) + Q(1, T, V)\lambda + Q(2, T, V)\lambda^2 + \ldots$$
$$= \frac{pV}{k_BT} = \sum_{N\geq 0}\left(Q_N \cdot e^{\frac{\mu N}{k_BT}}\right) = 1 + \sum_{N\geq 1}\left(Q_N \cdot e^{\frac{\mu N}{k_BT}}\right) \quad (5.10)$$

where the expansion coefficients correspond to the canonical partition functions, with $N - 1$ particles for the N-th term. Note that $Q_0(T, V) = Q(0, T, V) = 1$ because for $N = 0$, the system has only one possible state with zero energy.

Introducing the notation $z = \frac{Q_1\lambda}{V}$, Eq. 5.10 can be written as:

$$\Xi(V, T, \mu) = 1 + \sum_{N\geq 1}\left(\frac{Q_N V^N}{Q_1 N}\right) \cdot z^N \quad (5.11)$$

where $Z_N\,(V, T)$ is the classical configurational integral , defined as:

$$Z_N = \int_V e^{-W(q_i)} dq_1 \ldots dq_N \quad (5.12)$$

Here, $W\,(q_i)$ represents the intermolecular force potential, and q_i represents the coordinates of the particles in the volume V.

Taking the logarithm of both sides of Eq. 5.11, expanding the right-hand side, and dividing by ρ, we have:

$$\frac{P}{k_BT} = \sum_{j\geq 1} b_j\,(T)\, z^j \quad (5.13)$$

where $\rho = \sum_{j\geq 1} j b_j\,(T)\, z^j$. The coefficients b_j depend on the configurational integrals:

$$1!Vb_1 = Z_1 = V \quad (5.14)$$

$$2!Vb_2 = Z_2 - Z_1^2 \quad (5.15)$$

Thus, the virial coefficients expressed in the virial expansion are:

$$B_2 = -b_2 \quad (5.16)$$

$$B_3 = 4b_2^2 - 2b_3 \quad (5.17)$$

In this way, the B_n coefficients have been expressed in terms of b_j, and b_j in terms of Z_N, and Z_N in terms of Q_N. Specifically, to determine B_2, one only needs Q_1 and Q_2, and for B_3, Q_1, Q_2, and Q_3 are needed.

Focusing on the second virial coefficient B_2, which is related to two-body interactions, if we denote the interaction potential between two molecules separated by distance r as $W(r)$ (assuming spherically symmetric potential), then $B_2(T)$ can be expressed as:

$$B_2(T) = -\frac{1}{2}\int_0^{\infty}\left[e^{-\frac{W(r)}{k_BT}} - 1\right]4\pi r^2 dr \tag{5.18}$$

This expression allows us to derive the parameters of the interaction potential by measuring the values of the second virial coefficient. For complex functional forms of the potential (such as the Lennard-Jones potential), it is not possible to obtain a closed-form solution for the integral in Eq. 5.18. However, for potentials with simple functional forms, an explicit integral expression can be derived, which shows the dependence of the second virial coefficient on the parameters of the intermolecular potential.

For example, for a hard sphere potential 4.5, the expression for $B_2(T)$ is:

$$\begin{aligned} B_2(T) = 2\pi \cdot \mathcal{N} \cdot \left[\int_0^{\sigma}\left(1 - e^{-\infty/k_BT}\right)r^2 dr \right. \\ \left. + \int_{\sigma}^{\infty}\left(1 - e^{-0/k_BT}\right)r^2 dr\right] = \frac{2\pi \cdot \mathcal{N}\sigma^3}{3} \end{aligned} \tag{5.19}$$

Similarly, for a square-well potential 4.10, we have:

$$\begin{aligned} B_2(T) = 2\pi \cdot \mathcal{N} \cdot \left[\int_0^{\sigma}\left(1 - e^{-\infty/k_BT}\right)r^2 dr \right. \\ \left. + \int_{\sigma}^{k\sigma}\left(1 - e^{\varepsilon/k_BT}\right)r^2 dr + \int_{k\sigma}^{\infty}\left(1 - e^{-0/k_BT}\right)r^2 dr\right] = \\ = \frac{2\pi \cdot \mathcal{N}\sigma^3}{3} \cdot \left[1 - \left(\lambda^3 - 1\right) \cdot \left(e^{\varepsilon/k_BT} - 1\right)\right] \end{aligned} \tag{5.20}$$

These expressions explicitly show the dependence of the second virial coefficient on the parameters of the intermolecular potential, and they are typical applications of molecular thermodynamics, which allows deriving macroscopic volumetric properties of fluids in terms of molecular properties (such as the diameter of a hard sphere, σ).

5.4.1 Virial Equation of State for Osmotic Pressure

In the case of osmotic pressure for dilute liquid solutions, the extension of the McMillan-Mayer theory allows for the nearly unchanged extension of all results related to the relationship between the second virial coefficient and the interaction potential, as found for real gases.

To achieve this correspondence, it is necessary to choose an appropriate statistical ensemble that directly provides an expansion of osmotic pressure. The system described in Sect. 5.3 consists of two cells separated by a semipermeable membrane that allows only component 1 (solvent) to pass through but not component 2 (solute), which remains confined to one of the cells (cell II).

The most suitable statistical ensemble to describe this system is the semi-grand canonical ensemble, which is open with respect to one component (component 1) but closed with respect to the other (component 2).

By extending the grand canonical partition function to a two-component system (cell II), we have:

$$\Xi\left(\mu_1, \mu_2, T, V\right) = \sum_{N_1, N_2 \geq 0} Q_{N_1 N_2} \cdot e^{\frac{\mu_1 N_1}{k_B T}} e^{\frac{\mu_2 N_2}{k_B T}} = \tag{5.21}$$

$$= \sum_{N_2 \geq 0} \Psi_{N_2}\left(\mu_1, T, V\right) \cdot e^{\frac{\mu_2 N_2}{k_B T}} \tag{5.22}$$

where $\Psi_{N_2}\left(\mu_1, T, V\right)$ is defined as the semi-grand canonical partition function, which represents a sort of canonical partition function of component 2 in an environment consisting of component 1 at a constant chemical potential μ_1. Alternatively, in a more rigorous manner, it is the partition function of a system at fixed T and V, open to component 1 (μ_1 fixed) and closed with respect to component 2 (N_2 fixed).

In the particular case of $N_2 = 0$, we have:

$$\Psi_0\left(\mu_1, T, V\right) = \sum_{N_1 \geq 0} Q_{N_1} \cdot e^{\frac{\mu_1 N_1}{k_B T}} = e^{\frac{pV}{k_B T}} \tag{5.23}$$

which represents the grand canonical partition function of cell I.

For the virial expansion of osmotic pressure, we first need to define the ideal reference system, which is the condition of ideality for the liquid solution, preferably defined in a way that can be directly compared with the condition of a perfect gas, which is the starting point for deriving the virial series expansion of real gas pressure.

In the case of a gas, ideality is achieved at very low pressures, corresponding to very small gas densities, such that the interactions between particles can be neglected. In terms of gas fugacity (based on number density), in this case, we have $f_{gas} \rightarrow p_{gas}$ as $p_{gas} \rightarrow 0$.

Regarding colloidal solutions, the "parallel" ideality condition is infinite dilution of the solute within the solvent. That is, as $\rho_2 \rightarrow 0$ (ρ_2 being the solute number density), we can write van't Hoff's law (Eq. 1.9):

$$Z_2 = \frac{\pi}{\rho_2 k_B T} = 1 \tag{5.24}$$

where Z_2 is defined as the compressibility factor.

For nonideal colloidal mixtures (at concentrations ρ_2 of solute higher than those where interactions start to influence solution behavior), we can write the series expansion of osmotic pressure:

$$Z_2 = \frac{\pi}{\rho_2 k_B T} = 1 + \sum_{n \geq 2} B_n^* (\mu_1, T) \rho_2^n \tag{5.25}$$

In this case, the osmotic virial coefficients $B_n^* (\mu_1, T)$ depend not only on the solvent's chemical potential μ_1 but also on temperature since the interactions between solute molecules occur in the solvent medium. Similar to the virial coefficients for the virial expansion of hydrostatic pressure, the second osmotic virial coefficient is found to be:

$$B_2^* (\mu_1, T) = 2\pi \cdot \int_0^\infty \left[1 - e^{-\frac{W(r,\mu_1,T)}{k_B T}} \right] r^2 dr \tag{5.26}$$

where $B_2^* (\mu_1, T)$ represents the contribution to non-ideality due to the two-body interactions of solute molecule 2 with a solvent at chemical potential μ_1 and temperature T. Similarly, the interaction potential $W (r, \mu_1, T)$.[5] Similarly to what was seen in the previous paragraph, for the second osmotic virial coefficient, it is possible to derive an integral form that explicitly depends on the parameters of the intermolecular potential in the case that the latter assumes a simple form (hard sphere potential or square-well potential).

5.5 Virial Coefficients of the Osmotic Pressure in Classical Thermodynamics

A liquid solution containing a solute at concentration c is nonideal if the activity coefficients of water and the solute are different from unity. To describe the non-ideality of such systems, the activity is introduced for the i-th component:

$$a_i (\mathbf{x}) = \gamma_i (\mathbf{x}) \cdot x_i \tag{5.27}$$

[5] $W (r, \mu_1, T)$ represents the mean-field interaction potential (interactions between solute molecules 2 occur in the solvent medium corresponding to the solvent with chemical potential μ_1 and temperature T).

where $\mathbf{x}$ is the vector of mole fractions of the components present in the system. Based on this definition, it is intuitive to identify liquid solutions as nonideal when the corresponding activity coefficients deviate from unity. In other words, activity represents a measure of the actual influence of individual components, which depends on the concentration of the components (x_i) and the activity coefficient $\gamma_i(\mathbf{x})$, which corrects for the concentration effect. Ideal liquid solutions exhibit unity values for the activity coefficients.

From the definition of activity, it is also possible to introduce the osmotic coefficient of the i-th component, defined as:

$$\phi_i(\mathbf{x}) = \frac{\ln(a_i(\mathbf{x}))}{\ln(x_i)} = 1 + \frac{\ln(\gamma_i(\mathbf{x}))}{\ln(x_i)} \tag{5.28}$$

The water osmotic coefficient is usually determined using Raoult's reference (for $x_w \rightarrow 1 \quad \Rightarrow \gamma_w^R \rightarrow 1,\ \phi_w^R \rightarrow 1$):

$$\phi_w = \frac{\ln(a_w)}{\ln(x_w)} = 1 + \frac{\ln(\gamma_w)}{\ln(x_w)} \tag{5.29}$$

The mole fraction of water x_w is given by:

$$x_w = \frac{n_w}{n_w + n} = \frac{c_w}{c_w + c} \tag{5.30}$$

For very low solute concentrations, $c_T = c + c_w \simeq c_w = \rho_w$, where ρ_w is the molar concentration of water, which only depends on temperature (e.g., at 25 °C, there are 1000 g of water in 1 dm^3 of water, corresponding to $\frac{1000}{18} = 55.56$ moles; therefore, at 25 °C, the molar density of water is $\rho_w(25\,°\text{C}) = 55.56\,\frac{\text{moles}}{\text{L}}$). Therefore, the molar fraction of water in Eq. 5.30 reduces to:

$$x_w = 1 - \frac{c}{\rho_w} \tag{5.31}$$

which, when substituted into the general expression for the water osmotic coefficient:

$$\phi_w = -\frac{\ln(a_w)}{\ln\left(1 - \frac{c}{\rho_w}\right)} \tag{5.32}$$

for $\frac{c}{\rho_w} \rightarrow 0$, $\ln\left(1-\frac{c}{\rho_w}\right) \simeq -\left(\frac{c}{\rho_w}\right)$;[6] therefore, the expression for the water coefficient ϕ_w is:

$$\phi_w = -\frac{\rho_w \cdot \ln(a_w)}{c} \tag{5.33}$$

The coefficient ϕ_w is analogous to the compressibility factor z for hydrostatic pressure:

$$z = \frac{P}{\rho \cdot RT} \tag{5.34}$$

where ρ is the density of the gas; similarly, the relation that connects the osmotic pressure π to the coefficient ϕ_w is:

$$\phi_w = \frac{\pi}{c \cdot RT} \tag{5.35}$$

Therefore, the molar fraction of water in Eq. 5.30 reduces to:

$$x_w = 1 - \frac{c}{\rho_w} \tag{5.36}$$

which, when substituted into the general expression for the water coefficient:

$$\phi_w = -\frac{\ln(a_w)}{\ln\left(1-\frac{c}{\rho_w}\right)} \tag{5.37}$$

for $\frac{c}{\rho_w} \rightarrow 0$, $\ln\left(1-\frac{c}{\rho_w}\right) \simeq -\left(\frac{c}{\rho_w}\right)^7$; therefore, the expression for the water coefficient ϕ_w is:

$$\phi_w = -\frac{\rho_w \cdot \ln(a_w)}{c} \tag{5.38}$$

[6] This expression is derived from the McLaurin series expansion of the logarithmic function:

$$\ln\left(1-\frac{c}{\rho_w}\right) \simeq \ln\left(1-\frac{c}{\rho_w}\right)\Big|_{c=0} + \frac{\partial}{\partial c}\left[\ln\left(1-\frac{c}{\rho_w}\right)\right]_{c=0} \cdot c + \ldots = -\frac{c}{\rho_w}.$$

[7] This expression is derived from the McLaurin series expansion of the logarithmic function:

$$\ln\left(1-\frac{c}{\rho_w}\right) \simeq \ln\left(1-\frac{c}{\rho_w}\right)\Big|_{c=0} + \frac{\partial}{\partial c}\left[\ln\left(1-\frac{c}{\rho_w}\right)\right]_{c=0} \cdot c + \ldots = -\frac{c}{\rho_w}$$

The compressibility factor , as known, can be expanded in a series with respect to the gas density ρ:

$$z = 1 + \sum_{n=2}^{\infty} B_n(T) \cdot \rho^{n-1} \tag{5.39}$$

The coefficients $B_n(T)$ are called virial coefficients and can be expressed as:

$$B_n(T) = \frac{1}{(n-1)!} \cdot \left(\frac{\partial^{n-1} z}{\partial \rho^{n-1}} \right)_{\rho=0} \tag{5.40}$$

Similarly, by extension, the coefficient ϕ_w can be expanded in a series with respect to the solute concentration c:

$$\phi_w = \frac{\pi}{c \cdot RT} = 1 + \sum_{n=2}^{\infty} B_n(T) \cdot c^{n-1} \tag{5.41}$$

The coefficients $B_n(T)$ are called osmotic virial coefficients and can be defined, analogous to the case of osmotic coefficients, as:

$$B_n(T) = \frac{1}{(n-1)!} \cdot \left(\frac{\partial^{n-1} \phi_w}{\partial c^{n-1}} \right)_{c=0} \tag{5.42}$$

In particular, for the second virial coefficient, we can write:

$$B_2(T) = \left(\frac{\partial \phi_w}{\partial c} \right)_{c=0} \tag{5.43}$$

Using the expression for the water coefficient in Eq. 5.29, we find, for $B_n(T)$:[8]

$$B_2(T) = -\left[\frac{\partial}{\partial c} \left(\frac{\rho_w \cdot \ln(a_w)}{c} \right) \right]_{c=0} = -\rho_w \cdot \left[\frac{1}{c} \cdot \frac{\partial \ln(\gamma_w)}{\partial c} - \frac{\ln(\gamma_w)}{c^2} \right]_{c=0} \tag{5.44}$$

[8] Expanding the expression of the second virial coefficient:

$$B_2(T) = -\left[\frac{\partial}{\partial c} \left(\frac{\rho_w \cdot \ln(a_w)}{c} \right) \right]_{c=0} = -\rho_w \cdot \left[\frac{\partial}{\partial c} \left(\frac{\ln(a_w)}{c} \right) \right]_{c=0} =$$

$$= -\rho_w \cdot \left[\frac{\left(\frac{\partial \ln a_w}{\partial c} \right) \cdot c - \ln a_w}{c^2} \right]_{c=0} = -\rho_w \cdot \left[\frac{\left(\frac{\partial \ln(\gamma_w x_w)}{\partial c} \right) \cdot c - \ln(\gamma_w x_w)}{c^2} \right]_{c=0}$$

Furthermore, as seen earlier:

$$\ln x_w \simeq -\frac{c}{\rho_w} \rightarrow \frac{\partial \ln x_w}{\partial c} \simeq -\frac{1}{\rho_w}$$

The virial coefficients, as seen earlier, can be related to the parameters of the van der Waals equation of state , always taking into account their definition in terms of coefficients of the Maclaurin expansion of the compressibility factor z with respect to the molar density ρ, which, in the case of gases, corresponds to the inverse of specific volume $\rho = \frac{1}{v}$ and in the case of solutions, it is equal to the molar concentration of the solute $\rho = c$.

For example, for the second virial coefficient (osmotic, in the case of liquid solutions):

$$B_2(T) = \left(\frac{\partial z}{\partial \rho}\right)_{\rho=0} = \left(\frac{\partial z}{\partial c}\right)_{c=0}$$

If we consider the definition of the compressibility factor, for instance, through the van der Waals equation rewritten for osmotic systems:

$$z = \frac{1}{1 - b\,c} - \frac{a\,c}{RT}$$

We can derive:

$$B_2\,(T) = \lim_{c \to 0}\left(\frac{b}{(1 - bc)^2} - \frac{a}{RT}\right) = b - \frac{a}{RT}$$

This expression can be interpreted, also in light of the expression of B_2 obtained for the square-well potential , which includes, like the van der Waals model, a repulsive potential term (hard sphere) and an attractive potential term.

Therefore, knowing the relationship between the water activity coefficient γ_w and the solute concentration c, we can obtain the value of the virial coefficient.

The derived expression of the second coefficient for the square-well potential is (Eq. 5.20):

$$B_2\,(T) = \frac{2\pi\,\mathcal{N}\,\sigma^3}{3} \cdot \left[1 - \left(\lambda^3 - 1\right) \cdot \left(e^{\frac{\varepsilon}{k_B T}} - 1\right)\right]$$

---

which, when substituted into the previous expression:

$$B_2\,(T) = -\rho_w \cdot \left[\frac{1}{c} \cdot \left(\frac{\partial \ln \gamma_w}{\partial c} + \frac{\partial \ln x_w}{\partial c} - \frac{\ln \gamma_w}{c} - \frac{\ln x_w}{x}\right)\right]_{c=0} =$$

$$= -\rho_w \cdot \left[\frac{1}{c} \cdot \left(\frac{\partial \ln \gamma_w}{\partial c} - \frac{1}{\rho_w} - \frac{\ln \gamma_w}{c} + \frac{1}{\rho_w}\right)\right]_{c=0} =$$

$$= -\rho_w \cdot \left[\frac{1}{c} \cdot \frac{\partial \ln (\gamma_w)}{\partial c} - \frac{\ln (\gamma_w)}{c^2}\right]_{c=0}.$$

In the approximation of $1/T \rightarrow 0$ (which is valid in the vicinity of room temperature). In fact, the exponential term, dependent on temperature, can be approximated by its Maclaurin series expansion:

$$e^{\frac{\varepsilon}{k_B T}} \simeq \left(e^{\frac{\varepsilon}{k_B T}}\right)_{\frac{1}{T}=0} + \left(\frac{d\left(e^{\frac{\varepsilon}{k_B T}}\right)}{d\left(\frac{1}{T}\right)}\right)_{\frac{1}{T}=0} \cdot \frac{1}{T} = 1 + \frac{\varepsilon}{k_B} \cdot \frac{1}{T} \tag{5.45}$$

Using this expression, we obtain:

$$\begin{aligned} B_2(T) &= \frac{2\pi \mathcal{N} \sigma^3}{3} \cdot \left[1 - \left(\lambda^3 - 1\right) \cdot \frac{\varepsilon}{k_B} \cdot \frac{1}{T}\right] = \\ &= \frac{2\pi \mathcal{N} \sigma^3}{3} - \left[\frac{2\pi \mathcal{N} \sigma^3}{3} \cdot \left(\lambda^3 - 1\right) \cdot \frac{\varepsilon}{k_B}\right] \cdot \frac{1}{T} \end{aligned} \tag{5.46}$$

Thus, the following relationships between the van der Waals constants and the molecular parameters of the square-well potential are obtained:

The result regarding the covolume b is actually derived from the following consideration: the volume excluded by two spherical molecules of diameter σ is equal to the volume of a sphere with radius σ, which is $\frac{4\pi\sigma^3}{3}$. Therefore, the covolume associated with a single molecule is half of that, which is $\frac{2\pi\sigma^3}{3}$. Consequently, the volume excluded by one mole is exactly equal to the covolume by definition, and it corresponds to the result found in Eq. 5.46. If we examine the term related to the attractive contribution, however, we can infer that the product $b \cdot \left(\lambda^3 - 1\right)$ represents the term associated with the binary attractive interaction. In this region of space, the cumulative attractive energy for one mole will be precisely $\varepsilon \cdot \mathcal{N}$.

Additional Resources and Recommended Literature

A good reference for the volumetric properties of gases and liquids is provided in [18]; the application of molecular theories to derive volumetric properties of fluids in thermodynamics can be found in [16].

Biopolymer Ionization and Salting-Out

6

Remember, the Force will be with you always.
Obi-Wan Kenobi, Star Wars: Episode IV - A New Hope

G. Lucas

Abstract

Why it is important to know this material?

The equation of the state of gases and solutions is the basic determinant of the behavior of these systems. The parallelism between gas pressure and osmotic pressure in liquid solutions highlights a unitary approach that enables the description of gas mixtures and liquid phase solutions at the macroscopic and microscopic level in similar ways.

What is the key idea?

The equations of state for gases and liquid solutions provide an operational framework to define the thermodynamic properties of these systems and their phase equilibria.

What is necessary to know already?

Classical thermodynamics tools are essential to utilize state equations in the thermodynamic analysis of fluid systems.

6.1 Introduction

Proteins are copolymers with a highly diverse chemical nature. As seen in Sect. 6.3.4, amino acids can have very different characteristics. This variety is not only the basis for the three-dimensional structure that proteins assume in space but also for their ability to perform various functions under different environmental conditions, maintaining remarkable versatility and adaptability.

L. Di Paola, *Fundamentals of Molecular Bioengineering*,
https://doi.org/10.1007/978-3-031-42022-1_6

In particular, protein molecules have, compared to other biopolymers with similar molecular weight, high solubility, which is linked to the presence of polar and ionic amino acids. Thanks to these amino acids, proteins ionize in aqueous solutions, exhibiting a specific distribution of positive and negative charges localized on their molecular surface. This distribution also results in a net charge, either positive or negative, which is determined by the algebraic sum of the charges present.

The properties and functions of amino acids and proteins are determined by their ionization properties. Thus, to comprehend and anticipate the structural and functional traits of these systems, it is imperative to provide a comprehensive description of these equilibria.

6.2 Ionization Equilibria for Amino Acids, Proteins, and Peptides

The general structure of amino acids is shown in Fig. 3.1. The amino and carboxyl groups tend to dissociate as an acid and a base, respectively, ionizing in solution. Specifically, at neutral pH, both groups are present in ionized form as $-COO^-$ and $-NH_3^+$, respectively.

In general, the dissociation equilibrium of an acid HA can be described according to the following kinetic scheme:

$$HA \underset{K_A}{\rightleftharpoons} H^+ + A^- \tag{6.1}$$

The acid dissociation reaction at equilibrium is associated with an equilibrium constant K_A, which, in the case of a dilute system, is defined as:

$$K_A = \frac{[A^-] \cdot [H^+]}{[HA]} \tag{6.2}$$

The corresponding pK_A is:

$$pK_A = -\log_{10}[K_A] = pH - \log_{10}\frac{[A^-]}{[HA]} \tag{6.3}$$

From the previous definition, the Henderson-Hasselbalch equation can be derived:

$$pH = pK_A + \log_{10}\frac{[A^-]}{[HA]} = pK_A + \log_{10}\frac{[\text{conjugate base}]}{[\text{conjugate acid}]} \tag{6.4}$$

The pK_A, in other words, represents the pH at which the acid is half-dissociated $([A^-] = [HA])$.

Some amino acids have a residue that can also ionize (acidic residues such as aspartate and glutamate, and basic residues such as lysine and arginine). This ionization equilibrium overlaps with that of the amino and carboxyl groups.

According to van't Hoff's law, the dissociation equilibrium constant of the acid K_A is related to the standard Gibbs free energy of ionization:

$$\Delta G^0 = -RT \cdot \ln K_A \tag{6.5}$$

ΔG^0 represents the difference in free energy between the products (H^+ and A^-) and the reactant (HA) when all components are present at standard conditions (concentration of 1 M in aqueous solution, at the temperature T of the system). The free energy of ionization ΔG_{ion} under conditions different from standard is given by:

$$\Delta G_{ion} = \Delta G^0 + RT \cdot \ln([H^+] \cdot [A^-]/[HA]) \tag{6.6}$$

At equilibrium, the concentrations of species in solution are related by the equilibrium constant $\frac{[H^+]_e \cdot [A^-]_e}{[HA]_e} = K_A$; since $\Delta G^0 = -RT \cdot \ln K_A$, at equilibrium, $\Delta G_{ion} = 0$. If, at the same time as the ionization equilibrium, there is another phenomenon associated with a free energy ΔG_e, the corresponding total energy is given by:

$$\Delta G_T = \Delta G_e + \Delta G_{ion} = \Delta G^0 + \Delta G_e + RT \cdot \ln([H^+] \cdot [A^-]/[HA]) \tag{6.7}$$

In this case as well, at equilibrium, $\Delta G_T = 0$; the concentration of H^+ at which the acid is half-dissociated ($[HA] = [A^-]$) is given by:

$$\begin{aligned} \Delta G_T &= \Delta G^0 + \Delta G_e + RT \cdot \ln[H^+] + RT \cdot \ln([A^-]/[HA]) \\ &= \Delta G^0 + \Delta G_e + RT \cdot \ln[H^+]_{1/2} = 0 \end{aligned} \tag{6.8}$$

and we obtain:

$$[H^+]_{1/2} = e^{-(\Delta G^0 + \Delta G_e)/RT} \tag{6.9}$$

The apparent pK'_A, which represents the pH at which the acid is half-dissociated in the presence of a parallel phenomenon associated with the free energy ΔG_e, is given by:

$$pK'_A = -\log_{10}[H^+]_{1/2} = \frac{(\Delta G^0 + \Delta G_e)}{2.303 \cdot RT} \tag{6.10}$$

On the other hand, the intrinsic pK_A, which is measured in the absence of other phenomena beyond ionization, is given by:

$$pK_A = \frac{\Delta G^0}{2.303 \cdot RT} \tag{6.11}$$

By comparing Eqs. 6.10 and 6.11, we can derive the free energy ΔG_e of the process that occurs alongside ionization:

$$\Delta G_e = 2.303 \cdot RT \cdot (pK'_A - pK_A) \tag{6.12}$$

Therefore, measuring the difference ΔpK_A with and without the intervention of another phenomenon concurrent with ionization allows to measure the energy associated with the combined phenomenon under the specified conditions.

In general, each amino acid undergoes ionization of the carboxyl and amino groups, resulting in a dipolar ion (also known as a zwitterion) that is in equilibrium with the two corresponding undissociated forms according to the kinetic scheme:

$$NH_3^+ - CHR - COOH \overset{1}{\rightleftharpoons} NH_3^+ - CHR - COO^- \overset{2}{\rightleftharpoons} NH_2 - CHR - COO^- \tag{6.13}$$

This corresponds to the two pK values:

$$\begin{aligned} pK_1 &= p\left(\frac{[NH_3^+ - CHR - COO^-] \cdot [H^+]}{[NH_3^+ - CHR - COOH]}\right) \\ &= pH + \log\left(\frac{[NH_3^+ - CHR - COOH]}{[NH_3^+ - CHR - COO^-]}\right) \end{aligned} \tag{6.14}$$

$$\begin{aligned} pK_2 &= p\left(\frac{[NH_2 - CHR - COO^-] \cdot [H^+]}{[NH_3^+ - CHR - COO^-]}\right) \\ &= pH + \log\left(\frac{[NH_2 - CHR - COO^-]}{[NH_3^+ - CHR - COO^-]}\right) \end{aligned} \tag{6.15}$$

These values are listed for each amino acid.

For simplicity, denoting $A^{\pm}$, A^+, and A^- as the zwitterionic, cationic, and anionic forms of the amino acid, respectively, Eqs. 6.13 and 6.14 can be rewritten as:

$$pK_1 = p\left(\frac{[A^{\pm}] \cdot [H^+]}{[A^+]}\right) = pH - \log\left(\frac{[A^{\pm}]}{[A^+]}\right) \tag{6.16}$$

$$pK_2 = p\left(\frac{[A^-] \cdot [H^+]}{[A^{\pm}]}\right) = pH - \log\left(\frac{[A^-]}{[A^{\pm}]}\right) \tag{6.17}$$

Denoting $f^{\pm}$, f^{+}, and f^{-} as the fractions of dipolar, cationic, and anionic amino acid, respectively, from Eqs. 6.15 and 6.16, we have $f^{+} = \frac{[H^{+}]}{K_1} \cdot f^{\pm}$ and $f^{-} = \frac{K_2}{[H^{+}]} \cdot f^{\pm}$. Since $f^{+} + f^{-} + f^{\pm} = 1$, we find:

$$f^{\pm} = \frac{1}{1 + \frac{[H^{+}]}{K_1} + \frac{K_2}{[H^{+}]}} \tag{6.18}$$

$$f^{+} = \frac{K_1 \cdot [H^{+}]}{1 + \frac{[H^{+}]}{K_1} + \frac{K_2}{[H^{+}]}} \tag{6.19}$$

$$f^{-} = \frac{K_2/[H^{+}]}{1 + \frac{[H^{+}]}{K_1} + \frac{K_2}{[H^{+}]}} \tag{6.20}$$

The pH value at which $f^{\pm}$ is maximized corresponds to the *amino acid's isoelectric point*, which depends on the values of K_1 and K_2. By setting $\frac{df^{\pm}}{d[H^{+}]} = 0$ (maximum of the $f^{\pm}$ curve in Fig. 6.1), it is possible to find:

$$[H^{+}]p.i. = +\sqrt{K2 \cdot K_1} \tag{6.21}$$

Thus, the isoelectric point is given by $pI = p([H^{+}]p.i.) = \frac{1}{2} \cdot (pK1 + pK_2)$.

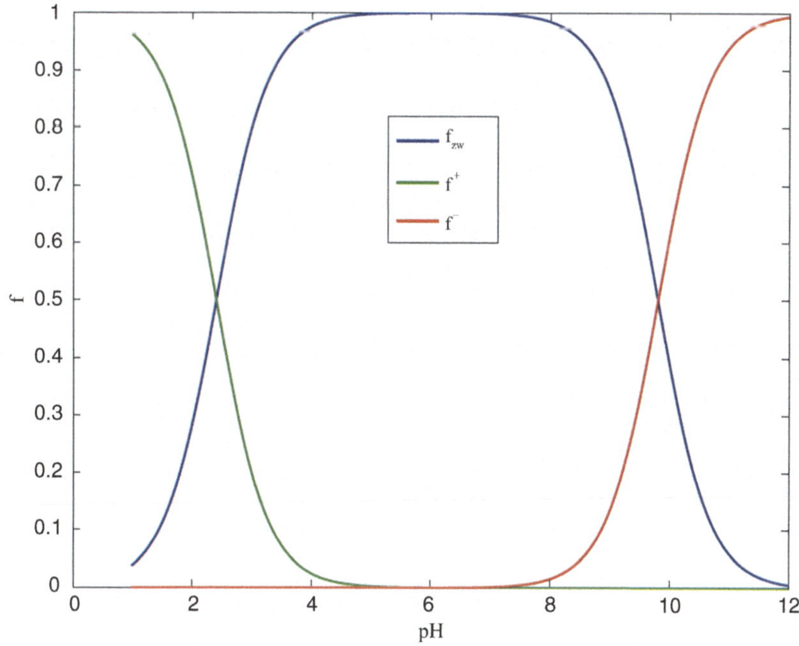

Fig. 6.1 Variation of the different ionic species resulting from the ionization of amino acids in aqueous solution as a function of pH

6.2.1 Ionization Equilibria of Residues in Peptides and Proteins

In polypeptides and proteins, amino acids lose their characteristic as dipolar ions because the amino and carboxyl groups are involved in peptide bond formation through condensation reactions. However, if ionizable residues are present, ionization equilibria still occur, although the pK_A values differ from those of free amino acids due to two effects:

1. Electrostatic interactions with nearby ionic or dipolar residues
2. Interference of ionization phenomena with the conformation equilibria of the polypeptide

In general, even considering only the ionization equilibria of the carboxyl and amino groups, it is observed that these equilibria are more favorable for free amino acids compared to polyamino acids.[1] Specifically, the values of pK_1 and pK_2 become closer as the size of the polyelectrolyte increases. Additionally, the interactions between the carboxylic and amino termini are associated with the difference between the pK values. In particular, by indicating ΔG_{ie}^0 as the free energy associated with the electrostatic interaction between the amino and carboxyl groups, it is obtained:

$$\Delta pK = (pK_1 - pK_2) = \frac{\Delta G_{ie}^0}{2.303 \cdot RT} \tag{6.22}$$

When amino acids polymerize to form chains of varying chemical composition (either uniform, as in polyamino acids, or nonuniform, as in polypeptides and proteins), the amino and carboxyl groups can no longer undergo ionization because they are involved in peptide bond formation. Only the two terminal groups, which are not involved in the bond formation, can still undergo ionization, but the extent of this ionization generally decreases as the length of the amino acid chain increases.

An example is given by the ionization equilibria of oligoalanine at different degrees of polymerization, shown in Table 6.1.

Table 6.1 Effect of the degree of polymerization of polyamino acids on the ionization equilibria of the amino and carboxyl groups for oligoalanine

	pK_1	pK_2
Ala (1)	2.34	9.69
Ala (2)	3.12	8.30
Ala (3)	3.39	8.03
Ala (4)	3.42	7.94

[1] Polyamino acids are defined as polymers resulting from the condensation of a single type of amino acid.

As the length of the oligopeptide increases, ionization becomes increasingly favorable. In fact, in a solution of a single amino acid, there is a strong attraction between the NH_3^+ and COO^- groups in the zwitterionic form at pH 7. This attractive interaction makes it difficult to remove a proton from the NH_3^+ group or add a proton to the COO^- group.

As the length of the polyamino acid increases, the attractive interactions between the terminal groups become weaker, and the corresponding ionization equilibria become increasingly "facilitated." As a result, they occur under less extreme pH conditions. This is why the pH range between pK_1 and pK_2 decreases with increasing degree of polymerization. According to Eq. 6.22, this corresponds to a decrease in the electrostatic interaction energy ΔG_{ie}^0 between the two terminal amino and carboxyl groups, resulting in greater ease of ionization for both groups.

When an amino acid residue within a protein chain undergoes ionization, which occurs within a well-defined three-dimensional structure, the ionization equilibrium occurs in an ionic environment determined by the local composition of the protein structure where the residue is present. It also depends on the conformation equilibria of the chain, which in turn depend on the ionization states of the residues along the chain.

It is, therefore, very complex to evaluate a priori the ionization state of amino acid residues and the corresponding *net surface charge*,[2] which is the electric charge equal to the algebraic sum of the charges on the surface of a protein molecule at a specific pH.[3]

The main effect of the net surface charge of a protein or, more generally, a biopolymer containing ionizable groups is to coordinate low molecular weight ions (mainly due to the electrolytic dissociation of salts) in the surrounding regions.

[2] As seen in Sect. 6.3.4, hydrophilic (polar and ionizable) residues tend to localize on the surface of the protein molecule. In other words, only some of the surface residues, those in direct contact with the aqueous solvent, are capable of ionization.

[3] The fact that the net surface charge of a protein is positive or negative indicates the prevalent nature of the ionic groups present on the protein molecule's surface, but it does not exclude the presence of charges of the opposite sign.

The simplest models for describing the electrostatic potential of a protein compound generally depict the protein molecule as a sphere with a radius equal to the molecular radius of the three-dimensional structure (spherical or spheroidal), with electric charge uniformly distributed over the entire protein surface.

In reality, ionized residues are distributed on the protein surface, including residues with opposite charges, as well as non-ionizable regions (polar or even hydrophobic). Thus, the distribution of surface charges is always extremely heterogeneous in reality.

Models that describe the electrostatic field generated uniformly provide sufficiently accurate descriptions for properties of solutions at moderate concentration (salting-out, electrical conductivity, electrophoretic mobility).

However, for all properties that can be interpreted through short-range interactions between charged surface regions (ion exchange chromatography, formation of protein-ion complexes), a more detailed description of the electrostatic field generated by the protein molecule is required, in terms of detailed spatial distributions of charges on the protein surface itself.

The final result of this coordination action is the so-called double ion layer (at greater distances, the coordination effect is negligible), which effectively produces a significant reduction in the initial charge density. Therefore, in the presence of salts, the ordering effect of microions around the charged surfaces of macroions leads to a significant reduction in the intensity of the electrostatic field generated by such macromolecules with an electrolytic surface nature.

As will be seen later, this phenomenon explains the behavior of solutions containing proteins and salts, especially regarding the variation of the physico-chemical properties of such systems following the addition of low molecular weight electrolytes (salts).

6.3 Salting-Out and Phase Equilibria of Solutions of Biomacromolecules

The term *salting-out* refers to the decrease in solubility of nonpolar compounds in an aqueous solution following the addition of a salt or, more generally, an electrolyte. In the context of protein systems, this phenomenon is used in protein separation processes, less frequently for purification operations. Additionally, the addition of salts and subsequent protein precipitation is frequently employed in laboratories to produce high-quality crystals for use in X-ray diffraction techniques, such as the study of the proteins' three-dimensional structure, as will be mentioned later.

To characterize the amount of charges present in a solution due to an electrolyte with concentration c_S, the ionic strength I is introduced:

$$I = \frac{1}{2} \cdot \sum_{i=1}^{n} (z_i^2 \cdot c_i) \tag{6.23}$$

where n is the number of ions resulting from the electrolytic dissociation of the salt, z_i is the charge of the i-th ion, and c_i is its concentration resulting from the dissociation of the electrolyte. This quantity is very useful for defining the total amount of charges introduced by an electrolyte, which depends not only on the salt concentration but also on the valence of the ions produced by dissociation.

Monoprotic ions, for example, have an ionic strength equal to the salt concentration itself, while for salts with polyprotic anions or cations, the ionic strength is greater than the salt concentration.

For instance, the ionic strength produced by a concentration c_S of $(NH_4)_2SO_4$ is:

$$I = \frac{1}{2} \cdot \left(c_{NH_4^+} \cdot (+1)^2 + c_{SO_4^{2-}} \cdot (-2)^2 \right) = \frac{1}{2} \cdot \left(2 \cdot c_S \cdot (+1)^2 + c_S \cdot (-2)^2 \right) = 3 \cdot c_S \tag{6.24}$$

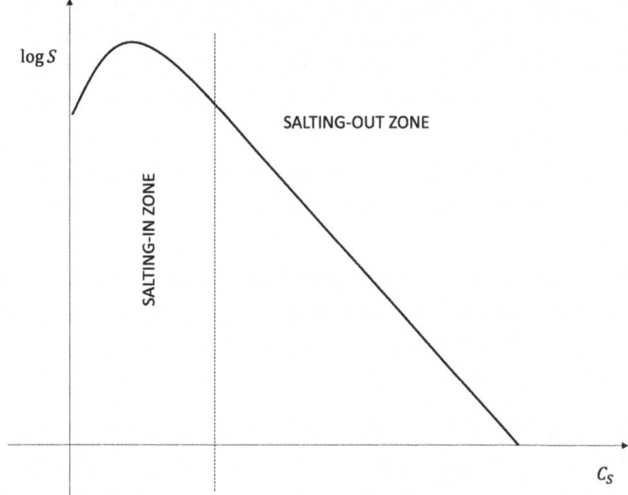

Fig. 6.2 Variation of the solubility of a protein compound with salt concentration: S represents the solubility of the biopolymer and c_S is the concentration of the added salt

Therefore, at the same concentration, ammonium sulfate produces an ionic strength three times that of sodium chloride for the same added salt concentration c_S.

A typical trend of the solubility of a protein, as a function of salt concentration, is shown in Fig. 6.2: at very low salt concentrations, the solubility of the protein increases, and then decreases, assuming a linear form in a semilogarithmic scale.

In the precipitation following salt addition, therefore, the ionic strength is used rather than the salt concentration since the extent of precipitation depends on the concentration of charges introduced through electrolyte dissolution. Proteins, and biopolymers in general, with surface charges, when dissolved in an aqueous solution, tend to be well-solubilized due to the repulsive electrostatic potential between like charges present on identical molecules, which prevents the formation of aggregates.

When an electrolyte is added to the solution containing the biopolymer, which can be seen in terms of electrostatic charges as a macroion with an assigned surface charge, the electrolyte dissociates, and the resulting microions localize around the macroion, forming an ion double layer that produces a significant decrease in the electrostatic field determined by the macroion's surface charge.

In other words, the charge of the macroion is screened by the presence of microions resulting from the electrolytic dissociation of the added salt. The formed ion double layer has a larger volume than that of the macroion, possessing a charge of the same sign as the macroion's net surface charge but of lesser magnitude.

This results in a significant reduction in the surface charge density of the macroion, consequently modifying all properties related to the net surface charge.

This screening effect of the macroion's surface charges is quantified by the *Debye-Hückel length*, defined as:

$$\lambda_D = \sqrt{\frac{\varepsilon_0 \varepsilon_r \cdot k_B T}{2N_{Av} \cdot e^2 \cdot I}} \tag{6.25}$$

where ε_0 and ε_r are the permittivity in vacuum and relative permittivity, respectively, referring to the dielectric medium constituted by the solution, e is the charge of an electron (equal to $-1.6 \cdot 10^{-19}$ C), and λ_D represents the range of influence of an ion's charge, depending on the presence of other ions in solution through the ionic strength I. At high ionic strengths, λ_D is small, indicating that the charge of the biopolymer is dampened in a reduced space due to the presence of other charges that provide stronger screening.

The effect of salting-in (i.e., the increase in solubility of the biopolymer due to the addition of salt), within a very limited range of ionic strength, is related to the fact that the addition of small amounts of electrolyte reduces the dielectric constant. This reduction leads to an increase in the repulsive electrostatic potentials between like charges on the surface of the macroions, resulting in a higher solubility of the macroions themselves, manifested as a decreased tendency to aggregate.

However, this positive effect on the solubility of the biopolymeric compound is quickly counterbalanced as the salt concentration increases, leading to the aforementioned screening effect.

6.3.1 Protein Salting-Out: The Hofmeister Series

The solubility data of proteins in the presence of salts are often analyzed using the empirical relationship known as the *Cohn-Edsall equation*:

$$\ln S = \beta - K_S \cdot I \tag{6.26}$$

Here, K_S is referred to as the salting-out constant, which depends on the specific biopolymer and salt used, while β is a constant that depends on the biopolymer and temperature but not on the salt. As seen in the figure, the equation represents the solubility of the protein at a hypothetical zero ionic strength.[4]

As mentioned above, not all ions behave in the same way regarding protein solubility. It has been observed that the most effective ions in precipitating protein compounds are those that, upon dissociation, produce monovalent cations and polyvalent anions (e.g., ammonium sulfate $(NH_4)_2SO_4$). For this reason, a scale

[4] It should be noted that in aqueous solution, even in the absence of added salts, there are always at least the ions H_3O^+ and OH^- resulting from the electrolytic dissociation of water.

of effectiveness in terms of salting-out, called the *Hofmeister series*, has been established. The series is ordered in decreasing order of effectiveness in precipitating protein compounds:

$$\text{citrate} > \text{sulfate} > \text{phosphate} > \text{chloride} > \text{nitrate} > \text{thiocyanate} \qquad (6.27)$$

By comparing the anions in Eq. 6.27 with Eq. 6.26, it is easy to see that the salts at the top of the Hofmeister series (i.e., those most effective in salting-out) are the ones with higher values of the salting-out constant K_S. In other words, the Hofmeister series can be seen as a decreasing series of salting-out constant values.

The effectiveness of salts in salting-out is also related to their ability to stabilize or disrupt the structure of water molecules in their vicinity. In this context, salts are referred to as *cosmotropic* when they have the capacity to strongly interact with the water structure. As a result, the water molecules near the ions exhibit a more ordered structure compared to the bulk water. The overall effect is the stabilization of the water structure, leading to a depletion of hydration water from the biopolymer molecules. This promotes their aggregation and precipitation. Another effect induced by these salts is the increased stability of the three-dimensional structure of biopolymers.

On the contrary, other ions with a small charge distributed over a large molecular surface tend to destabilize the water structure in their vicinity, favoring the hydration of biopolymer molecules. Such ions are called *chaotropic*, and they often interact with the structure of biopolymers, destabilizing them.

6.3.2 An Equation of State for Colloidal Solutions

Osmotic pressure measurements of protein aqueous solutions can also be used to derive the parameters of a corresponding equation of state that describes the behavior of the colloidal solution (water + protein). One method to obtain the equation of state in terms of parameters with clear molecular significance is based on the use of *perturbative theories*. These theories generally define thermodynamic quantities as the sum of two terms: one representing a reference molecular model (e.g., a system of rigid spheres) and the other term representing the perturbation from the reference conditions.

One such method is the *random-phase approximation (RPA)*, which, under dilute conditions, uses a reference system of rigid spheres. All additional interactions beyond the rigid sphere term are included in the perturbative term, assuming spherically symmetric interactions.

In detail, the thermodynamic quantities of interest are the specific excess Helmholtz free energy (per single molecule) A^{res} and the compressibility factor Z:

$$Z = \frac{\pi}{\rho k_B T} = \left(\frac{\pi}{\rho k_B T}\right)_{rif} + \frac{\rho U}{2k_B T} \tag{6.28}$$

$$\frac{A^{res}}{N_{Av} k_B T} = \left(\frac{A^{res}}{N_{Av} k_B T}\right)_{rif} + \frac{\rho U}{2k_B T} \tag{6.29}$$

where ρ is the number density of the macromolecule (number of macromolecules per unit volume of the solution) and U is the perturbation energy per single molecule, defined as:

$$U = 4\pi \int_V W(r) r^2 dr \tag{6.30}$$

where $W(r)$ is the mean-field interaction potential between pairs of macromolecules.

The reference term for the compressibility factor Z is obtained from the *Carnahan-Starling equation* valid for the osmotic pressure of solutes considered as hard spheres:

$$\left(\frac{\pi}{\rho k_B T}\right)_{rif} = \frac{1 + \eta + \eta^2 - \eta^3}{(1 - \eta)^3} \tag{6.31}$$

where $\eta = \left(\frac{\pi \cdot \rho}{6}\right) \sigma^3$ is the volumetric fraction referred to the spherical macromolecules of diameter σ.

Using the available expressions for the reference term of A^{res}, the chemical potential for the colloidal compound can be written as:

$$\frac{\mu - \mu'}{k_B T} = \frac{\eta\left(8 - 9\eta + 3\eta^2\right)}{(1 - \eta)^3} + \frac{\rho U}{k_B T} + \ln \rho \tag{6.32}$$

where $\frac{\mu'}{k_B T} = \log\left(\Lambda^3\right)$, $\Lambda = \frac{h}{\sqrt{2\pi M k_B T}}$ is the *de Broglie wavelength*, and M is the molecular weight of the macromolecular compound.

Starting from Eqs. 6.33, 6.31, and 6.32, it is possible to calculate the liquid/liquid phase equilibria of colloidal solutions using the following general expression related to the macromolecular compound:

$$\begin{cases} \mu^\alpha = \mu^\beta \\ \pi^\alpha = \pi^\beta \end{cases} \tag{6.33}$$

by applying Eqs. 6.28, 6.31, and 6.32.

A more detailed definition of the protein-protein interaction potential, which appears in the definition of the parameter U (Eq. 6.30) used in Eqs. 6.28 and 6.29. In general, for globular proteins (with spherical symmetry, where the potential is described along the single spatial coordinate r directed along the center-to-center axis of the molecules), it is possible to define an interaction potential between protein molecules as a sum of different contributions:

$$W_{PP}(r) = W_{SR}(r) + W_{EL}(r) + W_{DISP}(r) + W_{OSM}(r) + W_{ASS}(r) \tag{6.34}$$

where $W_{SR}(r)$ is the contribution of the hard sphere; $W_{EL}(r)$ is related to electrostatic interactions; $W_{DISP}(r)$ is related to dispersion forces (dipole-dipole, induced dipole-induced dipole, induced dipole-dipole); $W_{OSM}(r)$ is the attractive contribution of osmotic nature, present when polymers that bind to protein molecules are added; and finally, $W_{ASS}(r)$ is the strongly attractive potential related to highly specific associations, such as the formation of covalent bonds.

6.3.3 Phase Equilibria for Biomacromolecules

As seen in Chap. 1, chemical thermodynamics, through the definition of chemical potential, has as its main objective the analysis of phase equilibria. This study has practical applications in the development of systems and processes capable of separating components present in mixtures by forming phases with different compositions. In this sense, determining the interaction potentials between components in a mixture leads to the definition of chemical potentials that depend not only on the properties of the individual molecules but also on the interactions established in the mixture.

This type of application, which has allowed the development of traditional processes based on so-called unit operations, operations based precisely on the formation of two or more phases, can also be extended to the biotechnological world. This extension allows the development of processes and methods for separating biomolecules from mixtures of biological origin.

In particular, in the late 1990s, in parallel with the development of increasingly sophisticated methodologies for gene sequencing, methods and technologies were developed for large-scale production of biotechnological products for therapeutic use or in the field of fine chemistry (enzymes). In recent years, biotechnology has also found wide application in the environmental field due to the great ability of biological systems to eliminate pollutants through digestion.

All these applications involve the use of macromolecular or cellular systems that must be employed in specific formulations (phase, purity). Therefore, it is necessary to have predictive and suitable tools for designing production systems for this purpose based on the physical principles of phase formation in systems containing biomolecules.

Phase transitions within cells and tissues have been recognized as key mechanisms in controlling compartmentalization, i.e., the formation of separate regions (compartments) from a single uniform region.

For these reasons, it is necessary to introduce and extend the physical principles for determining phase equilibria to systems of biological nature. As will be shown below, the particular conformation, size, and physical properties of biological macromolecules result in peculiar phase equilibria that can be interpreted and predicted using the tools of molecular thermodynamics.

6.3.4 Molecular Thermodynamics of Solvation of Biomacromolecules

As seen previously, the process of solubilizing a protein molecule in a salt-containing solution can be divided into two steps:

1. Firstly, an adequate cavity must be formed to contain the protein molecule within the solution. This process requires an expenditure of energy, both enthalpic (breaking of water molecule bonds) and entropic (creation of the physical cavity).
2. Then, the molecule is placed inside the cavity and interacts with the water molecules it comes into contact with through its surface. This contribution is predominantly enthalpic, involving the interaction energies between the solute molecules and the solvent molecules that surround the inner surface of the cavity.

Based on these considerations, it is possible to derive the following expression relating solubility to ionic strength:

$$\ln S = x_0 + \beta_0 - (\Omega \cdot \sigma - \Lambda) \cdot I \tag{6.35}$$

Here, Ω is a constant dependent on the hydrophobic surface, σ is the increase in molal surface tension associated with the addition of electrolyte, β_0 is the *Debye-Huckel* term derived assuming the protein molecule has a spherical shape with a uniform charge (valid only at low ionic strengths), and Λ is the *Kirkwood* term, which considers the protein as a dipole. This contribution is proportional to the ionic strength. The constant x_0 includes all contributions except those related to cavity formation and electrostatic interactions between the protein and the electrolyte.

The functional forms of Eqs. 6.26 and 6.35 are the same. By comparison, we have $\beta = x_0 + \beta_0$ and $K_S = \Omega \cdot \sigma - \Lambda$. For example, a strong salting-out salt (high K_S) has a high value of σ, which is considered a direct measure of the ability of a salt to reduce the solubility of nonpolar compounds in aqueous solution.

Another way to interpret salting-out is through the preferential interactions of salts with protein compounds. Protein-salt interactions play a crucial role in the salting-out process. In particular, under salting-out conditions, salts are preferentially excluded from the region surrounding the protein.[5] Consequently, the salt concentration in this region is lower than in the bulk solution.[6] On the contrary, specific binding of ions to the protein molecule leads to an increase in protein solubility and, in some cases, conformation changes in the macromolecule.

From a thermodynamic perspective, salting-out can lead to phase equilibria in aqueous solutions containing proteins (or, more generally, biopolymers). The possibilities are numerous, depending on the ionic strength and pH conditions, and it is also possible to have the formation of metastable phases, which correspond to local minima, rather than absolute minima, of the Gibbs free energy associated with phase transitions.

Furthermore, regarding the occurrence of solid phases, they can assume different microscopic structures, ranging from amorphous to perfectly crystalline, depending on the ionic strength and, therefore, the magnitude of attractive interactions between protein molecules.

In general, the process of solubilizing a solute in a given solvent depends on solute-solute, solvent-solvent, and solute-solvent interactions. In particular, a solute is soluble in a given solvent if the Gibbs free energy associated with solute-solvent interactions ΔG_{solv} is sufficiently negative, meaning that the resulting interactions are attractive. In this case, the solvation process is associated with a significant decrease in Gibbs free energy.

On the contrary, a solute is insoluble in a given solvent if the Gibbs free energy associated with solute-solute and/or solvent-solvent interactions is more negative (stronger attractive interactions) than that associated with solute-solvent interactions.

This logics can be easily applied to the case of solubilizing protein compounds in an aqueous solvent, meaning that molecular and energetic parameters of macromolecules can be correlated with their behavior in solution. For example, it is reasonable to assume that the larger the protein molecule, the lower its solubility, while molecules with fewer surface charges are less soluble.

Additional Resources and Recommended Literature

A general definition of the ionization equilibrium of amino acids and proteins can be found in [14].

The salting-out of proteins described in terms of molecular thermodynamics of protein solutions is provided in [16].

[5] The preferential exclusion of salting-out salts from the spatial domain near the protein molecule is due to the unfavorable interactions that ions derived from the electrolyte dissociation establish with the hydrophobic regions of the protein surface.

[6] This type of measurement can be carried out through *equilibrium dialysis*, where the macromolecule is confined inside a bag whose walls consist of a semipermeable porous membrane that prevents the passage of the macromolecule to the external aqueous region.

7 Protein-Ligand Interactions and Binding

Thus light, in spite of what the crowd believe, if it be at an angle met by a dense body, is turned aside, reflected, and refracted, appearing different to different people.
Paradiso

D. Alighieri

Abstract

Why it is important to know this material?

Protein binding is implied in all biological mechanisms, and the definition of quantitative models and laws is essential to assess in a quantitative way the protein binding to small ligands, proteins, and nucleic acids.

What is the key idea?

Starting from molecular models, it is possible to determine the quantitative laws for protein binding.

What is necessary to know already?

It is essential to know how the protein structure is built and how it interacts with the environment, through a delicate interplay between stability and adaptability.

7.1 Introductive Elements

Most of the functions performed by protein compounds involve the formation of a complex between the protein molecule and other substances, either low or high molecular weight. In fact, protein molecules themselves serve little purpose, as they are always called upon to participate, either as actors (reactants or products) or as directors (enzymes), in a multitude of chemical reactions in which they need to interact with other molecules. The main classes of protein complexes are:

L. Di Paola, *Fundamentals of Molecular Bioengineering*,
https://doi.org/10.1007/978-3-031-42022-1_7

- **Protein-protein complexes**:
  - *Chaperonin*: Chaperonin molecules interact with unfolded or misfolded protein chains, mediating protein folding or contributing to structural stability.
  - *Protein aggregates*: The formation of aggregates among protein molecules, either of different species or identical ones, underlies many biologically relevant structures (e.g., corneal tissue, muscle tissue).
  - *Prions*: Misfolded proteins (prions) can interact with correctly structured molecules, inducing, often irreversibly, misfolding. This mechanism is the basis for degenerative diseases of the nervous tissue (such as Alzheimer's disease and bovine spongiform encephalopathy).
- **Protein-macromolecule complexes**:
  - *Protein-nucleic acids*: Gene regulation is known to depend closely on the specific interaction of nucleic acids with protein molecules, which act as carriers of the nucleic acids themselves, protecting them from potential unwanted interactions with the environment.
  - *Protein-polysaccharides*: The connective tissue is based on a dense network of glycosaminoglycans (belonging to the polysaccharide family), which is consolidated by interchain interactions mediated by protein molecules (collagen). These proteins confer specific chemical and physical characteristics to the system.
  - *Antibody-antigen*: An extremely important category of interactions between proteins and biopolymers is related to the action of antibodies, which exert their function by specifically interacting with other proteins or polysaccharides.
- **Protein-ligand complexes**:
  - *Albumin-hepatic toxins*: The interaction between serum albumin molecules in the plasma and hydrophobic hepatic toxins is essential for the excretory function of the liver.
  - *Hemoglobin-respiratory gases*: Hemoglobin binds respiratory gases (O_2, CO_2) and performs a dual function of supplying tissues with oxygen and removing waste substances through a complex mode of combined action, where the interaction with individual gases is conditioned by that with other components.
  - *Serum protein-drugs*: Pharmacologically active compounds have the ability to interact with serum proteins, forming complexes that facilitate their transport to target tissues and organs, as well as elimination through excretory pathways.
  - *Receptor-ligand*: In this case, the interaction mediates endocrine activity, acting as the basis for hormonal control of glandular functions.
- **Protein-surface complexes**:
  - *Adsorption on resins*: This process is the basis for the purification of compounds of various natures, where the immobilization of protein compounds on solid supports is due to attractive interactions between surface regions of the protein and the solid surface itself.
  - *Biocompatibility of materials*: The interaction between proteins and materials that come into contact with biological components underlies the definition of

biocompatibility. In particular, two protein molecules are used as reference in this definition: albumin and fibrinogen.

Albumin mediates the integration of the implanted material and its subsequent bioabsorption by the organism. Favorable interaction, in this perspective, indicates good biocompatibility of the material.

On the other hand, with regard to fibrinogen, an affinity of the material toward this compound indicates low biocompatibility of the material itself, as the adsorbed form of fibrinogen is also active in terms of participating in the coagulation cascade, leading to the formation of clots.

7.1.1 Simple Interaction: Single Class of Independent Binding Sites

Many biologically relevant phenomena involve interactions between macromolecules and molecules of lower molecular weight, called *ligands*: enzyme-substrate interactions, hormone-receptor interactions, ion-protein interactions, ion-nucleic acid interactions, etc.

To understand ligand-macromolecule equilibria, it is necessary to build models using the tools of classical and statistical thermodynamics for interpreting these equilibria.

The first approach consists of characterizing the system from both a macroscopic and microscopic perspective. In general, the association of ligands with multiple binding sites on macromolecules involves the definition of overall, macroscopic quantities and their corresponding microscopic counterparts. Specifically, let's consider the case of a macromolecule M containing n binding sites for ligand L, where each individual site has the same dissociation constant k. Furthermore, the sites are independent.[1] This means that the occupancy state of one site does not affect the dissociation constant k of the other sites. The equilibria can be expressed as follows:

$$M_0 + L \rightleftharpoons M_1 \tag{7.1}$$

$$M_1 + L \rightleftharpoons M_2 \tag{7.2}$$

$$\vdots$$

$$M_{i-1} + L \rightleftharpoons M_i \tag{7.3}$$

$$\vdots$$

$$M_{n-1} + L \rightleftharpoons M_n \tag{7.4}$$

[1] The assumption of identical and independent sites is usually valid when the protein exhibits a quaternary structure consisting of identical polypeptide subunits. Additionally, it is necessary that the binding site is not located near the bonds that hold the quaternary structure together.

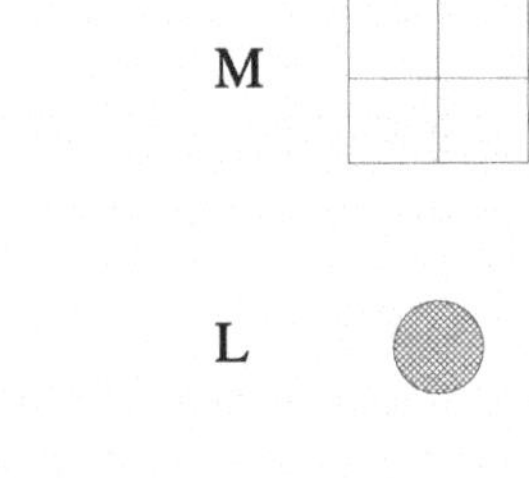

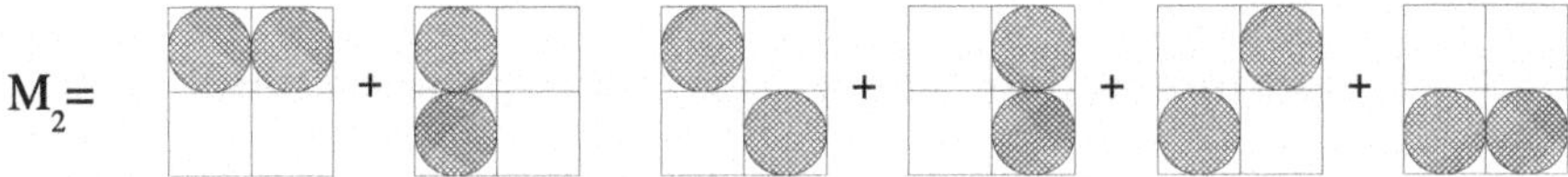

Fig. 7.1 Possible configurations for the arrangement of two ligand molecules on four available sites in a macromolecule

Here, M_i represents the macromolecule with i ligand molecules associated with i out of the n binding sites (where $i < n$). M_i encompasses all microscopic species that have i bound L molecules.

Example Let's consider the case of a macromolecule M with $n = 4$ available binding sites, of which two are occupied. The possible configurations for M_2 are shown in Fig. 7.1:

As observed, a total of six configurations are possible for the species M_2. Since all the binding sites have the same association constant k and are independent of each other, the concentrations of the six distinct species are equal, and the species themselves are indistinguishable from a macroscopic point of view. The number $\Omega_{n,i}$ represents the number of ways to place i ligands on n sites and is given by:

$$\Omega_{n,i} = \frac{n!}{(n-i)! \cdot i!} = \binom{n}{i} \tag{7.5}$$

Therefore, there exist $\Omega_{n,i}$ microscopic forms m_i corresponding to the form M_i:

$$[M_i] = \Omega_{n,i} \cdot [m_i] \tag{7.6}$$

Ligand-macromolecule equilibria are measured through the evaluation of a parameter ν representing the moles of ligand bound per mole of macromolecule. It is found that:

$$\nu = \frac{\sum\limits_{i=0}^{n} i \cdot [M_i]}{\sum\limits_{i=0}^{n} [M_i]} \tag{7.7}$$

The objective is to express ν in terms of the concentration of free ligand L. The generic macroscopic constant K_i is defined as:

$$K_i = \frac{[M_{i-1}] \cdot [L]}{[M_i]} \tag{7.8}$$

In this case, the constants differ depending on the degree of saturation of the biomolecule for statistical reasons. By substituting the relation 7.6, it is found that:

$$K_i = \frac{\Omega_{n,i-1} \cdot [m_{i-1}] \cdot [L]}{\Omega_{n,i} \cdot [m_i]} = \left(\frac{\Omega_{n,i-1}}{\Omega_{n,i}}\right) \cdot k \tag{7.9}$$

It is interesting to note that, even with a constant microscopic interaction, the macroscopic constants vary, and significantly so, with the degree of saturation. However, this does not imply that the degree of interaction modifies the interaction potential, as we will see later in the case of nonindependent sites. Using the recursive relations 7.8 and 7.9, we obtain:

$$[M_i] = \frac{[M_{i-1}] \cdot [L]}{K_i} = [M_0] \cdot \frac{[L]}{\prod_{j=1}^{i} K_j} \tag{7.10}$$

By introducing the relation 7.9 and rearranging, it is obtained:

$$\frac{\Omega_{n,i-1}}{\Omega_{n,i}} = \frac{i}{n-i+1} \tag{7.11}$$

which leads to:

$$\prod_{j=1}^{i} K_j = \frac{k^i}{\prod_{j=1}^{i} \frac{n-j+1}{j}} = \frac{k^i}{\binom{n}{i}} \tag{7.12}$$

and thus:

$$[M_i] = [M_0] \cdot \binom{n}{i} \cdot \left(\frac{[L]}{k}\right)^i \tag{7.13}$$

Substituting this expression into 7.7:

$$\nu = \frac{\sum_{i=1}^{n} i \cdot \binom{n}{i} \cdot \left(\frac{[L]}{k}\right)^i}{1 + \sum_{i=1}^{n} \binom{n}{i} \cdot \left(\frac{[L]}{k}\right)^i}, \tag{7.14}$$

the term in the denominator corresponds to the binomial expansion of $\left[1+\frac{[L]}{k}\right]^n$:

$$\left[1+\frac{[L]}{k}\right]^n=\sum_{i=1}^{n}\binom{n}{i}\cdot\left(\frac{[L]}{k}\right)^i \tag{7.15}$$

By differentiating both sides with respect to $\frac{[L]}{k}$ and multiplying by $\frac{[L]}{k}$, it is found:

$$n\cdot\left(\frac{[L]}{k}\right)\cdot\left(1+\frac{[L]}{k}\right)^{n-1}=\sum_{i=1}^{n}i\cdot\binom{n}{i}\cdot\left(\frac{[L]}{k}\right)^i \tag{7.16}$$

which is exactly the numerator in the definition of ν. Therefore, it is obtained:

$$\nu=\frac{n\cdot\left(\frac{[L]}{k}\right)\cdot\left(1+\frac{[L]}{k}\right)^{n-1}}{\left[1+\frac{[L]}{k}\right]^n}=\frac{n\cdot\left(\frac{[L]}{k}\right)}{1+\frac{[L]}{k}} \tag{7.17}$$

The desired expression for ν is obtained:

$$\nu=\frac{nL/k}{1+L/k} \tag{7.18}$$

which can also be written as:

$$\frac{\nu}{L}=\frac{n}{k}-\frac{\nu}{k} \tag{7.19}$$

This same expression can be derived very simply without statistical considerations. Let θ_i be the fraction of active sites for the i-th site (which, under the assumptions made, is indistinguishable from the remaining $n-1$ sites). It is obtained:

$$\theta_i=\frac{C_{i,b}}{C_{i,b}+C_{i,f}}=\frac{C_{i,f}\cdot(C_{i,b}/C_{i,f})}{C_{i,f}[1+(C_{i,b}/C_{i,f})]} \tag{7.20}$$

where $C_{i,b}$ represents the concentration of occupied i-th site and $C_{i,f}$ represents the concentration of the same site in the free state. Since $k=\frac{C_{i,f}\cdot L}{C_{i,b}}$, we have $\frac{C_{i,b}}{C_{i,f}}=L/k$, and therefore:

$$\theta_i=\frac{L/k}{1+L/k} \tag{7.21}$$

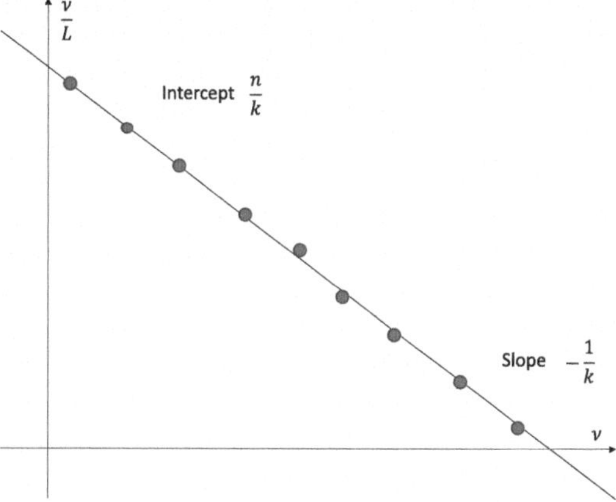

Fig. 7.2 Scatchard plot for a single class of sites with equal dissociation constant k, independent of the degree of site saturation

This same expression holds for all n sites, so we have:

$$\sum_{i=1}^{n} \theta_i = \nu = \sum_{i=1}^{n} \frac{L/k}{1 + L/k} = \frac{nL/k}{1 + L/k} \tag{7.22}$$

which is exactly the previously derived expression.

The expression $\frac{\nu}{L} = \frac{n}{k} - \frac{\nu}{k}$ can be used to study the relationship between ν and L for the simple case of independent sites. If the assumption of independent sites holds true, the graph is linear with an intercept on the y-axis equal to $\frac{n}{k}$ and on the x-axis at n, with a slope of $-\frac{1}{k}$. Therefore, through this graph, known as the *Scatchard plot*, it is possible to determine the equilibrium binding parameters n and k (Fig. 7.2).

In many cases, the parameter ν is more commonly replaced by the more general average saturation level y, defined as:

$$y = \frac{\text{number of bound sites}}{\text{total number of available sites}} = \frac{\sum_{i=0}^{n} i \cdot [M_i]}{n \cdot \sum_{i=0}^{n} [M_i]} = \frac{\nu}{n} \tag{7.23}$$

From 7.22, we can derive:

$$y = \frac{\frac{L}{k}}{1 + \frac{L}{k}} \tag{7.24}$$

This expression is equivalent to the *Langmuir isotherm* or the one obtained for a single binding site. In other words, when the sites are identical, the average saturation level does not differentiate between them and behaves as a function of the ligand concentration L in the same way as in the case of a single site on each individual biomolecule.

Another way to represent protein-ligand binding data is through the following kinetic scheme:

$$M + n\,L \overset{M}{\rightleftharpoons} L_n \tag{7.25}$$

This scheme describes the simultaneous occupation of all n available sites on each individual protein M.

In this case, it is possible to define a dissociation constant (macroscopic):

$$K_d = \frac{[M]\cdot[L]^n}{[ML_n]} \tag{7.26}$$

By introducing the average saturation level θ, we can write:[2]

$$\frac{\theta}{1-\theta} = \frac{[ML_n]}{[M]} = \frac{[L]^n}{K_d} \quad \rightarrow \quad \log\left(\frac{\theta}{1-\theta}\right) = n\cdot\log([L]) - K_d \tag{7.27}$$

Therefore, by plotting the data of $f(y) = \frac{y}{1-y}$ as a function of $[L]$ on a logarithmic scale, the slope provides the value of n, and the intercept gives the negative value of the dissociation constant K_d. This representation is known as *Hill plot* and is shown in Fig. 7.3.

An alternative method for representing protein-ligand interaction data, with reference to n available sites for binding, is to consider the formation of the i-th complex in a single step, as follows:

$$P + L \overset{K'1}{\rightleftharpoons} PL \;\; P + 2L \overset{K'2}{\rightleftharpoons} PL_2 \;\ldots\; P + nL \overset{K'n}{\rightleftharpoons} PLn$$

where the equilibrium constants are given by:

$$K_1' = \frac{[P]\cdot[L]}{[PL]} \;\; K'2 = \frac{[P]\cdot[L]^2}{[PL2]} = \frac{[P]\cdot[L]}{[PL]}\cdot\frac{[PL]\cdot[L]}{[PL_2]}$$

$$= K_1\cdot K_2\ldots K'i = \frac{[P]\cdot[L]^i}{[PLi]} = \prod_{j=1}^{i} K_j = \prod_{j=1}^{i}\frac{\Omega_{n,i-1}}{\Omega_{n,i}}\cdot k = \frac{k^i}{\binom{n}{i}}$$

where K_i are the macroscopic constants related to the Scatchard scheme.

[2] If n is the number of available sites on each individual macromolecule, then the fraction of occupied sites will simply be:

$$\theta = \frac{n\cdot[ML_n]}{n\cdot\{[M]+[ML_n]\}} = \frac{[ML_n]}{[M]_0}$$

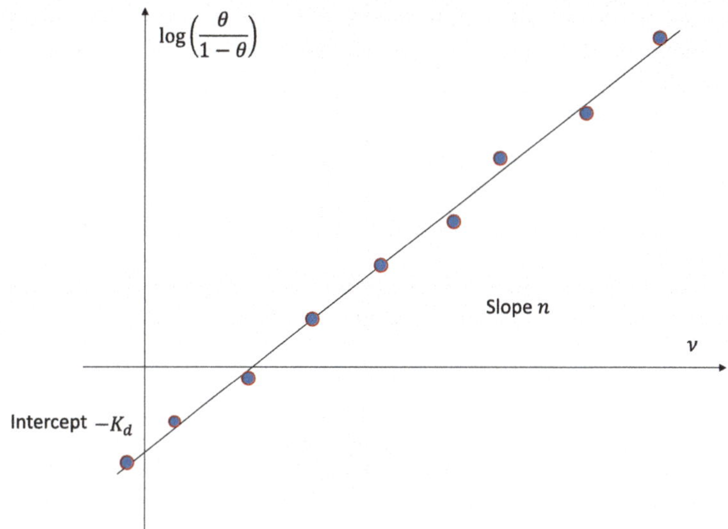

Fig. 7.3 Hill plot

If we consider the first scheme and denote $[P]b$ as the total concentration of bound protein and $[P]0$ as the total concentration of protein, and ϕ as the fraction of bound protein, we have:

$$\phi = \frac{[P]b}{[P]0} = \frac{[PL] + [PL_2] + \cdots + [PL_n]}{[P] + [PL] + [PL_2] + \cdots + [PL_n]} = \frac{\sum_{i=1}^{n} [P] \cdot \left(\frac{[L]}{k}\right)^i \cdot \binom{n}{i}}{\sum_{i=0}^{n} [P] \cdot \left(\frac{[L]}{k}\right)^i \cdot \binom{n}{i}}$$

$$= \frac{\sum_{i=1}^{n} \left(\frac{[L]}{k}\right)^i \cdot \binom{n}{i}}{1 + \sum_{i=0}^{n} \left(\frac{[L]}{k}\right)^i \cdot \binom{n}{i}} = 1 - \frac{1}{\left(1 + \frac{[L]}{k}\right)^n}$$

This expression is known as the *Adair equation*, which represents the case of independent and identical sites, identifiable with a single microscopic dissociation constant k, which is the same for all sites and independent of the degree of saturation itself ϕ.

In the case where two different ligands A and B interact with different affinities with the same site (for simplicity), two similar equations can be written for the binding equilibrium:

$$P + A \overset{K_A}{\rightleftharpoons} PA \tag{7.28}$$

$$P + B \overset{K_B}{\rightleftharpoons} PB \tag{7.29}$$

The association constants K_A and K_B can be expressed in terms of the concentrations in solution at thermodynamic equilibrium:

$$K_A = \frac{[P] \cdot [A]}{[PA]} \tag{7.30}$$

$$K_B = \frac{[P] \cdot [B]}{[PB]} \tag{7.31}$$

The concentration of free protein in solution can then be determined as:

$$[P] = \frac{K_A \cdot [PA]}{[A]} = \frac{K_B \cdot [PB]}{[B]} \tag{7.32}$$

By comparing the equations, it is possible to find:

$$\frac{[PA]}{[PB]} = \frac{K_B}{K_A} \cdot \frac{[A]}{[B]} \tag{7.33}$$

In the case where the concentrations of the two free ligands are the same ($\frac{[A]}{[B]} = 1$), the ratio between the concentrations of the two complexes is equal to the ratio of the two equilibrium constants:

$$\frac{[PA]}{[PB]} = \frac{K_B}{K_A} \tag{7.34}$$

The dissociation equilibrium constants K_A and K_B depend on temperature according to the van't Hoff equation:

$$\ln(K_A) = -\frac{\Delta G^0_{PA}}{RT} \tag{7.35}$$

$$\ln(K_B) = -\frac{\Delta G^0_{PB}}{RT} \tag{7.36}$$

Consequently, by applying the two previous expressions to Eq. 7.34, we obtain the temperature dependence of the ratio between the concentrations of the two complexes:

$$\frac{[PA]}{[PB]} = \frac{[A]}{[B]} \cdot \exp\left[-\frac{1}{RT} \cdot \left(\Delta G^0_{PB} - \Delta G^0_{PA}\right)\right] \quad (7.37)$$

Denoting $y_A = \frac{[PA]}{[A]}$ and $y_B = \frac{[PB]}{[B]}$ as the ratio between the complexed and free forms of the two ligands A and B, Eq. 7.37 can be rewritten as:

$$\frac{y_A}{y_B} = \exp\left[-\frac{1}{RT} \cdot \left(\Delta G^0_{PB} - \Delta G^0_{PA}\right)\right] \quad (7.38)$$

Or, in logarithmic form:

$$\log\left(\frac{y_A}{y_B}\right) = -\frac{1}{RT} \cdot \left(\Delta G^0_{PB} - \Delta G^0_{PA}\right) \quad (7.39)$$

Denoting $y_A = \frac{[PA]}{[A]}$ and $y_B = \frac{[PB]}{[B]}$ as the ratio between the complexed and free forms of the two ligands A and B, Eq. 7.37 can be rewritten as:

$$\frac{y_A}{y_B} = \exp\left[-\frac{1}{RT} \cdot \left(\Delta G^0_{PB} - \Delta G^0_{PA}\right)\right] \quad (7.40)$$

Or, in logarithmic form:

$$\log\left(\frac{y_A}{y_B}\right) = -\frac{1}{RT} \cdot \left(\Delta G^0_{PB} - \Delta G^0_{PA}\right) \quad (7.41)$$

Using this expression, it follows that if $\left(\Delta G^0_{PB} - \Delta G^0_{PA}\right) < 0$, i.e., if $\Delta G^0_{PB} < \Delta G^0_{PA}$, which means that ligand B has a higher affinity for protein P compared to ligand A, then as T increases, $\left(\frac{y_A}{y_B}\right)$ decreases, indicating a preference for the complex $[PB]$ over $[PA]$. Additionally, the slope of the line in Fig. 7.4 becomes steeper as the difference in affinity between the two ligands $-\left(\Delta G^0_{PB} - \Delta G^0_{PA}\right)$ increases.

Another interesting aspect related to the formation of protein-ligand complexes is that if the ligand exhibits higher affinity for the native form (N) rather than various denatured forms (U), it automatically stabilizes the native form, even without considering the possibility of forming energetically more stable complexes than the ligand-free native form.

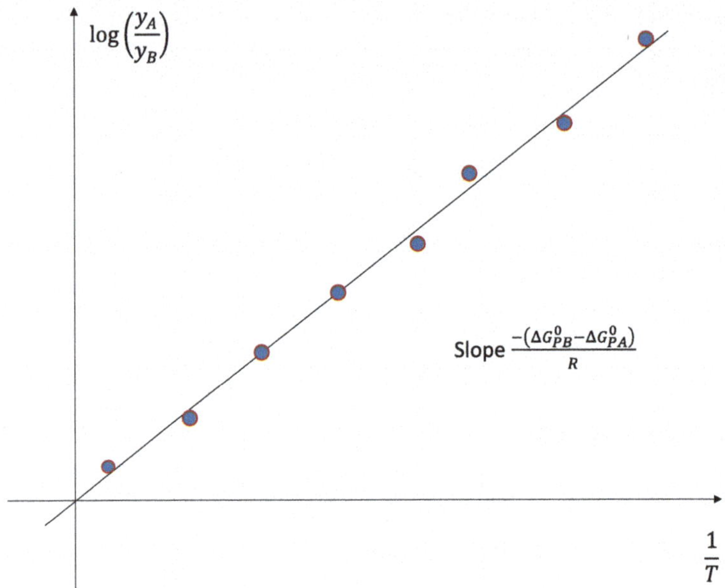

Fig. 7.4 Competitive binding of two ligands A and B to the same protein P; $y_A = \frac{[PA]}{[A]}$ and $y_B = \frac{[PB]}{[B]}$

Indeed, if the ligand preferably forms a more stable complex with the native form rather than the denatured forms, it means that in the presence of a denaturing agent, the reversible reactions occurring in solution are as follows:

$$N \underset{K_U}{\rightleftharpoons} U \tag{7.42}$$

$$N + L \underset{k}{\rightleftharpoons} NL \tag{7.43}$$

Here, K_U is the equilibrium constant for denaturation 7.42, and the equilibrium constants can be written in terms of the concentrations of the various species:

$$K_U = \frac{[U]}{[N]} \tag{7.44}$$

$$k = \frac{[N] \cdot [L]}{[NL]} \tag{7.45}$$

Consequently, the total concentration of protein in the free and complexed form is given by:

$$[N]_T = [N] + [NL] \tag{7.46}$$

The apparent denaturation equilibrium constant, which takes into account all native forms present in solution, is given by:

$$K_U^{app} = \frac{[U]}{[N]_T} = \frac{[U]}{[N]+[NL]} \\ = K_U \cdot \left(\frac{k}{k+[L]}\right) \quad (7.47)$$

As can be seen, the apparent denaturation constant K_U^{app}, in the presence of ligand binding reaction with ligand L, is smaller than the denaturation constant K_U, and the multiplicative factor less than one depends on the ligand concentration and the dissociation equilibrium constant k.

7.1.2 Multiple Sites, Allosteric Interactions, and Competitive Binding

A variety of phenomena can cause deviation from the linear shape of the Scatchard plot. These phenomena can be described by relaxing the assumptions made for constructing the Scatchard plot. The first case we consider is the presence of multiple classes of independent sites. In this case, the modified form of the Scatchard plot obtained, for example, with n_1 moles of sites with microscopic dissociation constant k_1, n_2 with constant k_2, etc., is given by:

$$\nu = \sum_{i=1}^{m} \frac{n_i L/k_i}{1+L/k_i} \quad (7.48)$$

From which we can easily find:

$$\frac{\nu}{L} = \sum_{i=1}^{m} \frac{n_i/k_i}{1+L/k_i} \quad (7.49)$$

This type of expression can be used to determine the binding parameters for different types of binding. Let's consider the case of only two classes of sites, i.e., n_1 binding sites with dissociation constant k_1 and n_2 sites with k_2. Equation 7.49 becomes:

$$\frac{\nu}{L} = \frac{n_1/k_1}{1+L/k_1} + \frac{n_2/k_2}{1+L/k_2} \quad (7.50)$$

Se si rappresenta questa espressione in funzione di L, si ottiene un grafico curvo:

In the case of two distinct classes of sites with n_1 sites having a dissociation constant k_1 and n_2 sites with k_2, the intercepts of this graph can be used to determine the binding parameters of the two classes of sites. As L approaches 0, $\frac{\nu}{L}$ has an indeterminate form $\frac{0}{0}$, and the value of this indeterminate form can be found using l'Hôpital's rule:

$$\lim_{L \to 0} \frac{d\nu}{dL} = \frac{n_1}{k_1} + \frac{n_2}{k_2} \tag{7.51}$$

There is also the physical constraint that $\frac{\nu}{L} = 0$ when $L = n_1 + n_2$.

A graph of $\frac{\nu}{L}$ as a function of L may not be linear even in the absence of multiple classes of sites with different affinities for the ligand. Another possible explanation is that the binding of a ligand to a specific site can influence the dissociation constant $k(\nu)$ of the other sites.

This case falls into the category of phenomena known as *allostery*, where the interaction between a ligand and a protein is modified by the binding of an effector molecule to an allosteric site (which is different from the active site). Specifically, activators of allosteric effectors intensify the activity of the protein, while allosteric inhibitors reduce the activation of the protein.

If the allosteric inhibitor is a product of the enzymatic reaction, the allosteric reaction can function as negative feedback control (*feedback inhibition*).

Furthermore, the allosteric activator can be the ligand itself (referred to as *homotropic allosteric modulator*), which is typically an activator. Alternatively, it can be a molecule different from the specific ligand (referred to as *heterotropic allosteric modulator*).

Let's consider the case of identical sites that are not independent from each other. In this case, the dissociation constant of the sites depends on the saturation level, ν. We assume that all sites initially have the same affinity for the ligand, meaning that the binding of the first ligand molecule to any site is equiprobable.

If we consider a macromolecule M with a single class of sites, we denote k_0 as the dissociation constant at $\nu = 0$. As ν increases, there is a variation in k. We define ΔG^0 as the relative standard free energy change for the dissociation of a bound ligand, given by:

$$\Delta G^0 = \Delta G_0^0 + RT\phi(\nu) \tag{7.52}$$

Here, ΔG_0^0 represents the standard free energy change for the dissociation process at $\nu \to 0$. We can also express $\Delta G_0^0 = -RT \ln k_0$ and $\Delta G^0 = -RT \ln k$. Substituting these expressions into Eq. 7.52, it is obtained:

$$k(\nu) = k_0 \cdot \exp(-\phi(\nu)) \tag{7.53}$$

Here, $\phi(\nu)$ is a function that depends on the effects of interactions between sites, which vary with the degree of saturation. Specifically, if $\phi(\nu)$ is a decreasing function in ν, then $k(\nu)$ is an increasing function in ν. This means that as the sites become occupied, the affinity of the ligand for the macromolecule decreases, indicating an enhanced contribution of repulsive interactions between the ligand and the site compared to attractive interactions. On the contrary, if $\phi(\nu)$ is an increasing function in ν, then $k(\nu)$ is a decreasing function in ν. This implies that as the binding sites become saturated, they exhibit an increasing affinity for the ligands themselves. This type of behavior falls under the broader class of *cooperative processes and interactions*. In particular, if binding increases affinity, it is referred to as *positive cooperativity*, whereas if it decreases affinity, it is called *negative cooperativity*.

Cooperativity can also be demonstrated using different models. For instance, by revisiting the Adair model, cooperativity can be observed when the dissociation constants of successive order complexes are different. For example, if the value of K_i decreases as the value of i increases, it is referred to as positive cooperativity. Conversely, if the value of K_i increases with increasing i, it is called negative cooperativity.

As an example, Fig. 7.5 shows curves representing the fraction of bound sites in the case of positive and negative cooperativity, compared to the case of independent and identical sites with a dissociation constant equal to the first dissociation value. The scenario pertains to the presence of two interacting sites.

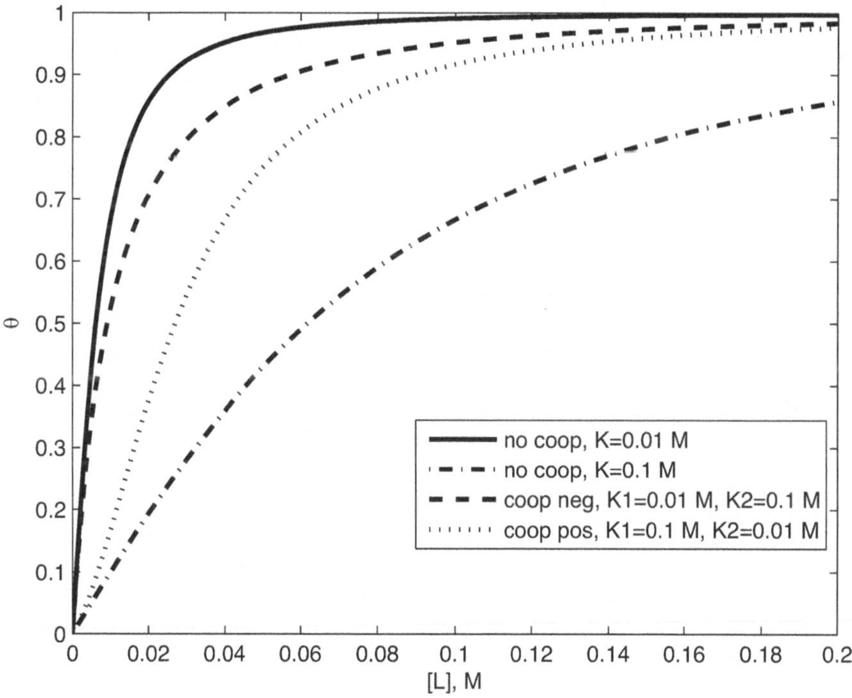

Fig. 7.5 Graph obtained from the Adair model for two sites, in the case of positive and negative cooperativity

It can be observed that in the case of positive cooperativity, at the same ligand concentration $[L]$, there is a higher protein-ligand affinity when the sites are not identical. The opposite behavior is observed in the case of negative cooperativity.

Furthermore, the sigmoidal shape is evident in cases of cooperativity, with an upward concavity for positive cooperativity and a downward concavity for negative cooperativity.

Additionally, interesting examples of interactions between macromolecules and ligands include the binding between plasma proteins and drugs and the behavior of allosteric proteins.

The bindings that occur between specific drugs and certain plasma proteins are crucial in determining the effectiveness and duration of action of a drug. In fact, only the free, unbound form of the drug exhibits specific pharmacological activity. It is therefore important, during the initial stages of experimentation and development of a pharmaceutical remedy, to determine the extent of this binding, that is, to identify which proteins are capable of "capturing" the drug molecule, the number of binding sites involved, and their respective affinities.

7.2 Molecular Models for Allostery

Several molecular models have been proposed, which in recent years have also utilized spectroscopy and NMR data for the analysis and validation of specific molecular mechanisms.

However, two molecular models proposed between the 1960s and 1970s still play a central interpretative role. They were formulated to describe the behavior of particular carrier molecules (hemoglobin) and the allosteric regulation of enzymatic activity.

The *concerted (or symmetric) model*, also known as the *Monod-Wyman-Changeux (MWC) model*, assumes the presence of two distinct states of the protein: T (*taut*, tense) and R (relaxed); the R conformation has a higher affinity for ligands, while the T conformation is more abundant in the absence of ligand.

In the MWC model, simplifying assumptions are made to evaluate the average saturation level:

- The quaternary structure is composed of identical chains (protomers) organized in a structure with at least one axis of symmetry.
- Each protomer contains a binding site for the ligand under consideration.
- There is a conformation equilibrium between two different states that exhibit different affinities for the specific ligand of interest.
- The affinity depends only on the conformation state and not on the saturation level of other sites, meaning that the formation of the protein-ligand complex does not induce changes in the tertiary structure.

The basis of this model is that there are two distinct forms of the protein that exhibit the same affinity for the individual ligand as the saturation level varies.

In other words, in terms of macroscopic constants, for n protomers and individual binding sites, there are two values of the dissociation constant k_R and k_T, respectively, for the R and T forms, which are identical for all sites in each form, regardless of the number of occupied sites in that particular form.

Similar to the case of identical and independent sites, in this case, the corresponding macroscopic constants K_{Ri} and K_{Ti} for the two forms depend on the statistical distribution of ligands in the sites and therefore vary with the degree of saturation.

The association equilibria are overlapped by an equilibrium between the two forms, which relates to the free form of the conformations.

In detail, the kinetic scheme is as follows:

$$T_0 \underset{K_C}{\rightleftharpoons} R_0 \tag{7.54}$$

$$\begin{array}{cc} R_0 + L \overset{K_{R1}}{\rightleftharpoons} R_1 & T_0 + L \overset{K_{T1}}{\rightleftharpoons} T_1 \\ \cdots & \cdots \\ R_{i-1} + L \overset{K_{Ri}}{\rightleftharpoons} R_i & T_{i-1} + L \overset{K_{Ti}}{\rightleftharpoons} T_i \\ \cdots & \cdots \end{array} \tag{7.55}$$

In this scheme, the macroscopic constants K_C (related to the conformation equilibrium of the free forms), K_{Ri}, and K_{Ti} are indicated. The different macroscopic dissociation constants are distinguished because, for statistical reasons, they vary with the degree of saturation, while the corresponding microscopic constants k_R and k_T remain constant. In fact, using the statistical derivation introduced for the Scatchard model:

$$K_{Ri} = k_R \cdot \frac{\Omega_{n,i-1}}{\Omega_{n,i}} = \frac{[L] \cdot [R_{i-1}]}{[R_i]} \tag{7.56}$$

$$K_{Ti} = k_T \cdot \frac{\Omega_{n,i-1}}{\Omega_{n,i}} = \frac{[L] \cdot [T_{i-1}]}{[T_i]} \tag{7.57}$$

The degree of saturation, referred to the sites, is given by:

$$y = \frac{R_1 + 2 \cdot R_2 + \cdots + n \cdot R_n + T_1 + 2 \cdot T_2 + \cdots + n \cdot T_n}{n \cdot (R_1 + R_2 + \cdots + R_n + T_1 + T_2 + \cdots + T_n)} = \frac{\sum_{i=0}^{n} i \cdot (R_i + T_i)}{\sum_{i=0}^{n} (R_i + T_i)} \tag{7.58}$$

Using a derivation similar to that of the Scatchard equation, the following expression for υ can be obtained:[3]

$$\nu = \frac{nL/k}{1 + L/k} \tag{7.61}$$

also expressed as:

$$\frac{\nu}{L} = \frac{n}{k} - \frac{\nu}{k} \tag{7.62}$$

---

[3] Derivation:

$$\sum_i [R_i] = [R_0] \cdot \sum_i \binom{n}{i} \cdot \left(\frac{[L]}{k_R}\right)^i = [R_0] \cdot \left(1 + \frac{[L]}{k_R}\right)^n$$

$$\sum_i [T_i] = [T_0] \cdot \sum_i \binom{n}{i} \cdot \left(\frac{[L]}{k_T}\right)^i = [T_0] \cdot \left(1 + \frac{[L]}{k_T}\right)^n$$

Introducing $\alpha = \frac{[L]}{k_R}$ and $\beta = \frac{k_R}{k_T}$, the two equations become:

$$\sum_i [R_i] = [R_0] \cdot (1 + \alpha)^n$$

$$\sum_i [T_i] = K_C \cdot [R_0] \cdot [1 + \beta\,\alpha]^n$$

Regarding the terms in the numerator of the equation, it is recognized that:

$$\sum_i i \cdot [R_i] = [R_0] \sum_i i \cdot \binom{n}{i} \alpha^i = [R_0]\, \alpha \cdot \frac{d}{d\alpha}\left(\sum_i \binom{n}{i} \alpha^i\right)$$

$$= [R_0]\, \alpha \cdot \frac{d}{d\alpha}\left([1 + \alpha]^n\right) = n\, [R_0] \cdot \alpha \cdot (1 + \alpha)^{n-1}$$

$$\sum_i i \cdot [T_i] = [T_0] \cdot \sum_i i \cdot \binom{n}{i} \cdot (\alpha\beta)^i = K_C \cdot [R_0] \cdot \alpha\beta \frac{d}{d\alpha}\left(\binom{n}{i} \cdot (\alpha\beta)^i\right)$$

$$= K_C \cdot [R_0] \cdot \alpha\beta \frac{d}{d\alpha}\left([1 + \beta\,\alpha]^n\right) \tag{7.59}$$

$$= n \cdot K_C \cdot [R_0] \cdot \alpha\beta \cdot [1 + \beta\,\alpha]^{n-1} \tag{7.60}$$

Substituting these expressions into the definition of y gives the desired general result for n protomers:

$$y = \frac{\alpha \cdot (1 + \alpha)^{n-1} + K_C \cdot \alpha \cdot \beta \cdot (1 + \beta\,\alpha)^{n-1}}{(1 + \alpha)^n + K_C\,(1 + \beta\,\alpha)^n}$$

where $\alpha = \frac{[L]}{k_R}$ and $\beta = \frac{k_R}{k_T}$.

If $K_C = 0$, we obtain the expression 7.24, which holds in the case of no conformation changes upon binding. Another notable result is obtained when $\beta \simeq 0$ (negligible affinity of the T form compared to the R form) and $K_C \gg 1$. Several molecular models have been proposed, which in recent years have also used spectroscopy and NMR data for the analysis and validation of specific molecular mechanisms.

The second model, called *sequential* or *KNF* (Koshland, Nemethy, Filmer) model, is based on the assumption that the bound and the unbound forms are present with different conformations, that is, the binding produces a conformation change for the protein molecule. The conformation changes upon binding occur in a different subunit of the protein chain from the active site. Even if the subunits can change conformations in an independent way, the change in one subunit favors the change in other units, reducing the energy required (see Fig. 7.6).

In other words, when a ligand binds to a specific subunit, it induces a shape modification in the protein, creating a thermodynamically favorable environment for other subunits to transition into a high-affinity state (*positive cooperativity*).

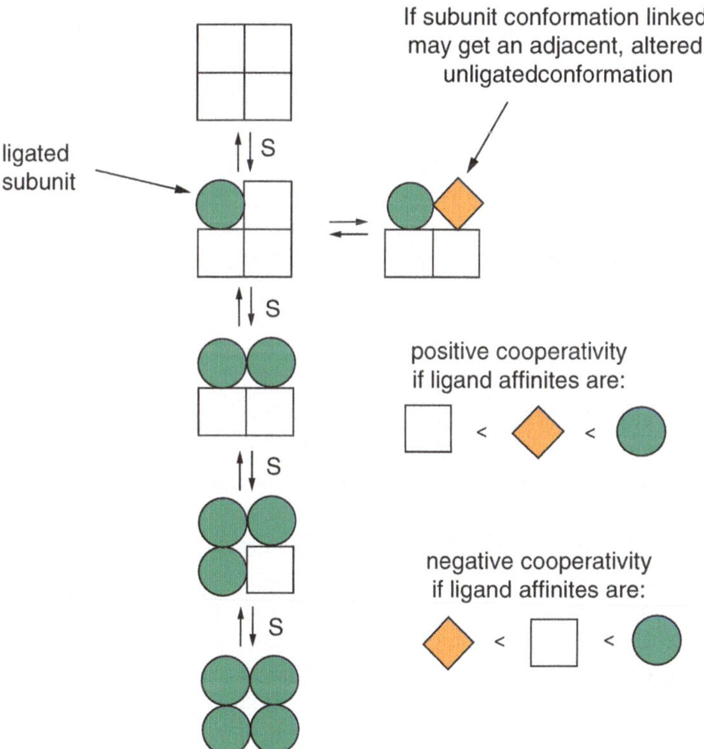

Fig. 7.6 Pictorial scheme for the KNF model

This conformation change can lead to *negative cooperativity*, where the subsequent binding site exhibits a decreased affinity for the ligand. This feature sets the KNF model apart from the MWC model, which proposes solely positive cooperativity.

The two proposed molecular models, even if proposed in the 1960s and 1970s, still play a significant interpretative role. They were formulated to describe the behavior of specific carrier molecules (hemoglobin) and the allosteric regulation of enzymatic activity.

7.2.1 The Allosteric Regulation of the Blood Hemoglobin

The allosteric regulation of the hemoglobin molecule provides a prominent example of the central role played by allostery in biological processes. The selection of hemoglobin as an illustrative model for understanding the key aspects of allostery stems from the extensive body of literature and research dedicated to this protein. Notably, hemoglobin was one of the first proteins for which the quaternary structure was resolved. Indeed, the Monod-Changeux-Wyman (MCW) model was developed to elucidate the functional behavior of hemoglobin and remained the predominant model for explaining the allosteric nature of protein molecules for a considerable period.

In hemoglobin occur all possible cases of interaction with multiple ligands and allosteric interaction between binding sites belonging to different or identical classes. The forms of hemoglobin are numerous, varying across species and even within the same species. In the case of the human form, for example, two forms are known: the adult physiological form and the fetal physiological form, known as Gower's form. There are also various pathological forms, including the one associated with sickle cell anemia. The protein has a quaternary structure, consisting of four domains held together by strong interactions. These domains are referred to as α_1, β_1, α_2, and β_2.

As shown in Fig. 3.6, the basic unit of hemoglobin is the $\alpha\beta$ protomer, and the two protomers are related by a binary axis of symmetry. The two subunits, α and β, have a very similar structure, practically overlapping each other, and they share many similarities with the structure of myoglobin, despite having low sequence homology. As shown in the figure, the α and β subunits are arranged at the vertices of a tetrahedron. The $\alpha_1\beta_1$ interface and its counterpart $\alpha_2\beta_2$ consist of approximately 30 residues and are sufficiently stable to resist mild denaturing treatments. On the other hand, the $\alpha_1\beta_2$ and $\alpha_2\beta_1$ interfaces involve 19 residues. The bonds between residues at the interface are predominantly hydrophobic in nature, but hydrogen bonds and ionic interactions are also present.

The specific function of hemoglobin in the body is the exchange of respiratory gases: the transport of oxygen to the tissues and the removal of carbon dioxide, which is transported and exchanged back with oxygen in the pulmonary alveoli. The site responsible for binding oxygen is the heme group (Fig. 3.6), present in each of the four subunits of the protein.

From a chemical point of view, hemoglobin heme is a porphyrin consisting of four pyrrole rings arranged in a plane, with an iron atom bound to it (see Fig. 3.6. The iron has six coordination bonds, four of which are involved in bonding with the nitrogen atoms of the porphyrin ring. In this electron configuration, when not bound to oxygen, the iron is in the ferrous state, Fe^{2+}. In the structure of hemoglobin, the heme group is located in a pocket, which reduces its reactivity compared to the free form that is not bound to a polypeptide chain. However, this positioning also reduces its interaction with carbon monoxide, preventing the deactivation of the hemoglobin molecule and increasing its functional stability as a result.

The interaction of hemoglobin with oxygen is a typical example of *homotropic allosteric interaction*. When a molecule of O_2 binds to the heme group present on one of the subunits, it changes the relative position of all the subunits, thereby altering the accessibility and reactivity of the remaining free heme groups located on the other units. The reactivity of the heme groups in the subunits comprising hemoglobin is comparable to that of the equivalent complex found in myoglobin. However, in myoglobin, the monomeric corresponding form, the allosteric behavior mentioned earlier, is completely absent. In fact, the allosteric behavior of hemoglobin, compared to that of myoglobin, is primarily attributed to the interactions between the monomeric subunits. The initial structural information for constructing a model of allosteric interaction in hemoglobin is based on the observation that it exists in two forms, known as the T (tense) and R (relaxed) states, corresponding to low and high affinity of hemoglobin for oxygen, respectively.

The deoxygenated form of hemoglobin, known as *deoxyhemoglobin*, corresponds to thc low-affinity T state. What is observed is that oxygen binding also promotes the formation of the R state, thereby further favoring the binding of additional oxygen molecules. This behavior is well-reflected in the sigmoidal saturation curve of hemoglobin. This curve, commonly used in medicine to represent the degree of oxygen saturation of hemoglobin, indicates the percentage of hemoglobin sites that are still free and available to bind new oxygen molecules. In contrast, the saturation curve of myoglobin exhibits a behavior consistent with a Scatchard plot in the absence of allosteric interactions, meaning its affinity does not vary with the degree of molecule saturation.

This behavior can be described by the Hill equation (Eq. 7.27), where the slope n is a direct measure of binding cooperativity. The T-to-R transition is associated with an increase in slope, reflecting the enhanced accessibility of the heme groups. This difference further confirms that the allosteric behavior of hemoglobin is related to the relative movement of the myoglobin subunits in the quaternary structure. When oxygen binds to the heme group of one subunit, it induces an opening of the overall protein structure, making the sites on other subunits more easily accessible.

7.3 Molecular Models of Enzymatic Reactions

A characteristic and widely applicable example of protein-molecule interactions in biology is represented by enzymatic reactions, which are based on the highly specific interaction between a protein compound called *enzyme* and a molecule called *substrate*.

Enzymes are biological catalysts, meaning they enhance the kinetics of specific biological reactions while remaining unchanged in concentration and form.

The nomenclature of enzymes, although not officially standardized, is often related to their function. Additionally, the suffix "-ase" is added to the root derived from the substrate's name. For example, urease is the enzyme that catalyzes the decomposition of urea, and alcohol dehydrogenase catalyzes the oxidative dehydrogenation reaction of an alcohol. However, some enzymes, such as trypsin and pepsin (proteolytic enzymes), do not fall under this classification.

The effect of enzymes on the energetics of the reactions they catalyze is not to modify the free energy of the substrate and reactants, i.e., the ΔG^0 of the enzymatic reaction, but rather to modify the activation energy of the reaction itself (see Fig. 7.7).

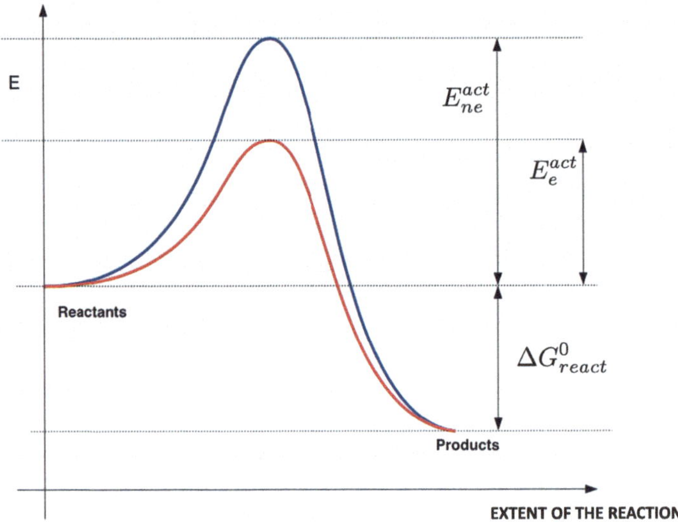

Fig. 7.7 Energetics of catalytic (enzymatic) reactions: the red line represents the reaction pathway in the presence of a catalyst, while the blue line represents the same reaction pathway in the absence of a catalyst. $\Delta G^0_{\text{reaz}} = G^0_{\text{products}} - G^0_{\text{reactants}}$ is the change in Gibbs free energy associated with the reaction, $E^{\text{att}}_{\text{ne}}$ is the activation energy in the absence of a catalyst, and $E^{\text{att}}_{\text{e}}$ is the same quantity relative to the catalyzed reaction

Table 7.1 Main differences between biological catalysts (enzymes) and non-biological catalysts

Enzymes	Non-biological catalysts
Specific	Nonspecific
Act only in the presence of cofactors	Do not require activation factors
Fragile 3D structures (easily degraded)	Greater resistance to deactivation
Maximum catalytic activity at T_{amb}	Maximum catalytic activity at elevated temperatures ($> T_{amb}$)

Catalytic reactions transform a specific reactant, called a substrate, into a product through an irreversible reaction, which can be schematized as follows:

$$S \overset{E}{\rightleftharpoons} P \tag{7.63}$$

The main differences between enzymes and non-biological catalysts are summarized in Table 7.1.

In particular, it is notable that biological catalysts require appropriate activation factors (cofactors) to function properly. This need is related to the high activity and specificity of these catalysts, which are designed to work in well-defined regions of the organism (organ or tissue targeted by enzymatic action). Outside of these regions, it is important for them to remain in an inactive state.

Cofactors are non-peptide compounds that chemically bind to the inactive form of the enzyme (called *apoenzyme*) to form a catalytically active complex, known as a *holoenzyme* or simply an enzyme.

Cofactors can be metallic ions (e.g., calcium ion Ca^{2+} serves as a cofactor for amylase, collagenase, and lipase) or organic in nature (in which case they are called coenzymes and are more commonly known as vitamins). The main difference between cofactors and *prosthetic groups* is that the latter establish an irreversible bond with the protein compound, while the controllable nature of enzymatic activity is linked to the formation of reversible and environment-dependent bonds between the cofactor and the apoenzyme.

When characterizing an enzyme, its source, i.e., the species of the originating organism, is always indicated. This is because enzymes that perform the same function in different organisms operate in different environments and exhibit structural and functional differences (catalytic activity).

The activity of an enzyme is measured in terms of the *turnover number* of the protein, which indicates the number of substrate molecules that react per catalytic site per unit of time.

In general, temperature has a contrasting effect: in a range near room temperature, increasing the temperature enhances catalytic activity, while at higher temperatures, conformation changes associated with thermal denaturation can abruptly reduce catalytic activity.

The activity of an enzyme is often modulated by the presence of low molecular weight molecules, which can also be the products of the reaction itself. If they act as cofactors, a systemic positive feedback control loop is established, whereas if they act as inhibitors, a negative feedback loop is formed.

The action of enzymes has been demonstrated through various experimental techniques to involve the formation of a substrate-enzyme complex. The complex formation reaction occurs at a specific active site. In general, the interactions between the substrate molecule and the active site on the enzyme are attractive in nature, and the bonds formed are primarily non-covalent.

As mentioned earlier, the initial approach to interpreting the specificity of the enzyme-substrate interaction considered a rigid chemical-structural recognition (*lock-and-key model*). However, this model proved insufficient in explaining the conservation of catalytic activity even after structural changes in the protein molecule. Therefore, this view has been modified to include the possibility of conformation adaptations of the protein structure following the formation of the enzyme-substrate bond (*induced fit model*), which can also occur due to local deformations of the specific interaction site (*hand-in-glove model*).

In terms of energy, the enzyme-substrate complex is more stable than the free enzyme, meaning that the enzyme-substrate complex corresponds to a deeper conformation potential well than the free form.

Finally, the mechanism by which the enzyme facilitates the reaction is based on two main phenomena:

(1) The enzyme can accelerate the reaction simply by bringing two substrate molecules together in an optimal position for the reaction to occur (energetic contribution of an entropic nature). (2) The substrate-enzyme binding can occur if the substrate is in the transition state. Thus, the state is stabilized or enforced by the bond between the enzyme and the substrate (contribution mainly of an enthalpic nature).

7.3.1 Mechanism of Enzymatic Activity

As previously discussed, enzymatic reactions can be summarized by an irreversible reaction scheme that transforms the substrate into products (Eq. 7.63). The disappearance rate ($-r_S$) represents the number of substrate moles consumed per unit volume and time.

The molecular mechanism of this reaction can be described by the following sequence of reactions:

$$S + E \underset{k_{-1}}{\overset{k_1}{\rightleftharpoons}} ES \tag{7.64}$$

$$ES \rightharpoonup k_2 E + P \tag{7.65}$$

Here, ES refers to the enzyme-substrate complex, which is in equilibrium with the free enzyme and substrate. If the reaction is elementary,[4] the following equilibrium can be written for the first reaction:

$$k_1 \cdot c_S \cdot c_E = k_{-1} \cdot c_{ES} \tag{7.66}$$

Thus, we can write:

$$K = \frac{c_S \cdot c_E}{c_{SE}} = \frac{k_{-1}}{k_1} \tag{7.67}$$

where K is the dissociation constant of the enzyme-substrate complex. Similarly, for the decomposition of the complex according to the reaction scheme of the second reaction, assuming irreversible reaction conditions, we can write the rate of product formation as:

$$r_P = \frac{dc_P}{dt} = k_2 \cdot c_{ES} = (-r_S) \tag{7.68}$$

Furthermore, based on the second reaction scheme, we can write:

$$(-r_S) = k_2 \cdot c_{ES} = -\frac{dc_S}{dt} \tag{7.69}$$

The rate of complex formation, combining the two reactions, is given by:

$$r_{ES} = k_1 \cdot c_S \cdot c_E - k_{-1} \cdot c_{ES} - k_2 \cdot c_{ES} = k_1 \cdot c_S \cdot c_E - (k_{-1} + k_2) \cdot c_{ES} = \frac{dc_{ES}}{dt} \tag{7.70}$$

Additionally, we can express the following balance equation for the enzyme:

$$c_E^0 = c_E + c_{ES} \tag{7.71}$$

Therefore, to determine the four unknowns c_S, c_E, c_{ES}, and c_P, it is necessary to solve the three ordinary differential equations 7.68)–(7.70 with the following boundary conditions:

$$\begin{cases} c_S(0) = c_S^0 \\ c_E(0) = c_E^0 \\ c_{ES}(0) = 0 \\ c_P(0) = 0 \end{cases} \tag{7.72}$$

[4] Elementary reactions are those in which the reaction orders for individual components correspond to their stoichiometric coefficients.

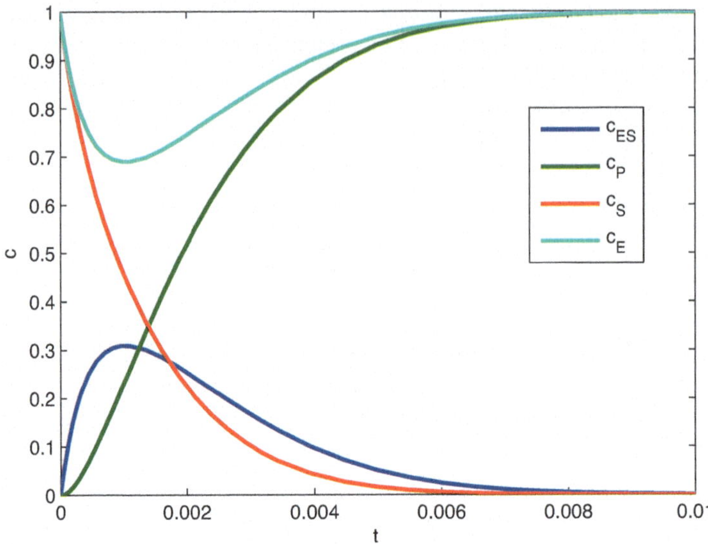

Fig. 7.8 Time evolution of the concentrations of free enzyme, enzyme-substrate complex, substrate, and product

An example of the trends obtained by numerically solving the system of ordinary differential equations representing the time evolution of concentrations c_E, c_S, c_{ES}, and c_P is shown in Fig. 7.8.

By assuming the quasi-steady-state approximation for the enzyme-substrate complex, i.e., assuming that the rate of complex formation is identically zero at every instant, from Eq. 7.71 we obtain:

$$\frac{dc_{ES}}{dt} = 0 \Rightarrow \quad k_1 \cdot c_S \cdot c_E = (k_{-1} + k_2) \cdot c_{ES} \tag{7.73}$$

which leads to:

$$c_{ES} = \frac{k_1 \cdot c_S \cdot c_E}{k_{-1} + k_2} \tag{7.74}$$

From Eq. 7.71, we have:

$$c_{ES} = \frac{k_1 c_S \cdot \left(c_E^0 - c_{ES}\right)}{k_{-1} + k_2} \tag{7.75}$$

and solving for c_{ES} gives:

$$c_{ES} = \frac{k_1 \cdot c_S \cdot c_E^0}{k_{-1} + k_2 + k_1 \cdot c_S} \tag{7.76}$$

By substituting this expression into Eq. 7.69, we obtain the substrate consumption rate:

$$(-r_S) = k_2 \cdot c_{ES} = \frac{k_1 \cdot k_2 c_E^0 \cdot c_S}{(k_{-1} + k_2) + k_1 c_S} = \frac{k_2 c_E^0 \cdot c_S}{\frac{(k_{-1}+k_2)}{k_1} + c_S} \tag{7.77}$$

Introducing the parameters $r_{MAX} = k_2 c_E^0$ and $K_M = \frac{(k_{-1}+k_2)}{k_1}$, representing the *maximum reaction rate* and the *Michaelis-Menten constant*, respectively, Eq. 7.77 can be simplified as follows (*Michaelis-Menten kinetic equation*):

$$(-r_S) = \frac{r_{MAX} \cdot c_S}{K_M + c_S} \tag{7.78}$$

Figure 7.9 illustrates the variation of the Michaelis-Menten reaction rate as a function of substrate concentration.

As seen previously, enzymatic reactions can be summarized by an irreversible reaction scheme for the transformation of substrate into products (Eq. 7.63); the disappearance rate $(-r_S)$ indicates the number of substrate moles consumed per unit volume and time.

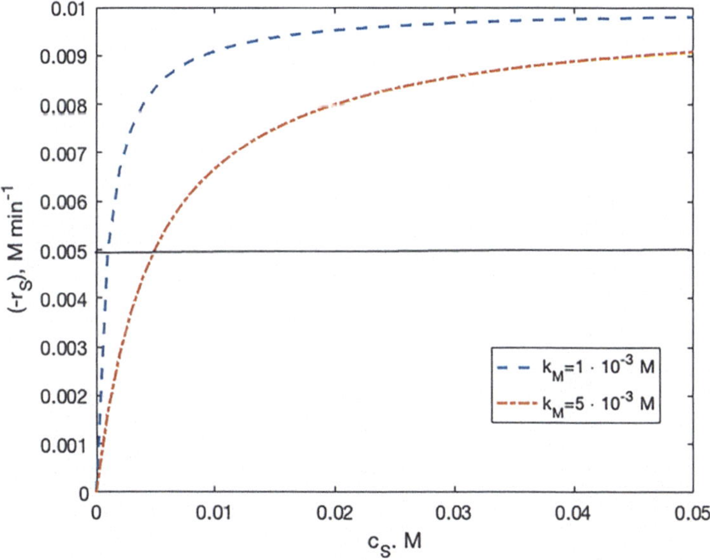

Fig. 7.9 Michaelis-Menten kinetics: the two curves have the same asymptotic value, equal to $r_{MAX} = 1 \cdot 10^{-2}\ M\ min^{-1}$, and different values of K_M

7.3.2 Enzyme Inhibition

7.3.2.1 Competitive Inhibition

Many drugs are enzyme inhibitors (consider commonly used antiretrovirals for AIDS treatment) and are characterized based on their specificity (ability to bind specifically to a single target enzyme molecule) and their efficacy, which refers to the concentration required to reduce enzymatic activity.

In general, the bond established between the inhibitor and the enzyme molecule can be reversible or irreversible. In the former case, the potential for intermolecular interaction is not very intense and has a long range (hydrophobic interactions, van der Waals forces, or electrostatic interactions), while in the case of irreversible binding, a reaction occurs (different from the one leading to product formation), resulting in the formation of a stable complex that is deactivated with respect to the reaction with the substrate; the intermolecular forces are very intense and short range, producing a covalent bond.

Below (Fig. 7.10), are described the most common mechanisms of reversible inhibition, which occur individually in biological systems or in more complex combinations, forming the basis of biological process regulation. As for irreversible inhibition, the scheme followed is specific to the enzyme-inhibitor pair and cannot be generalized.

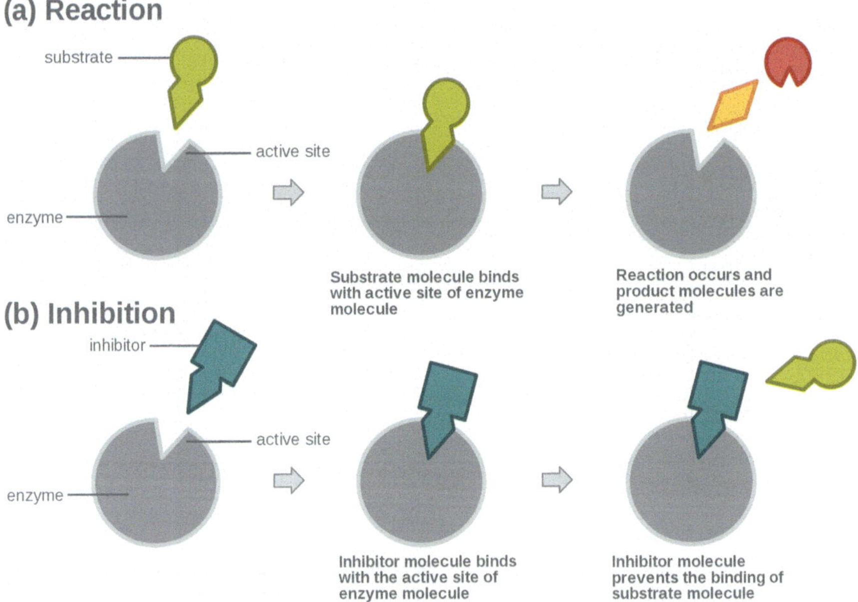

Fig. 7.10 Molecular scheme for competitive inhibition

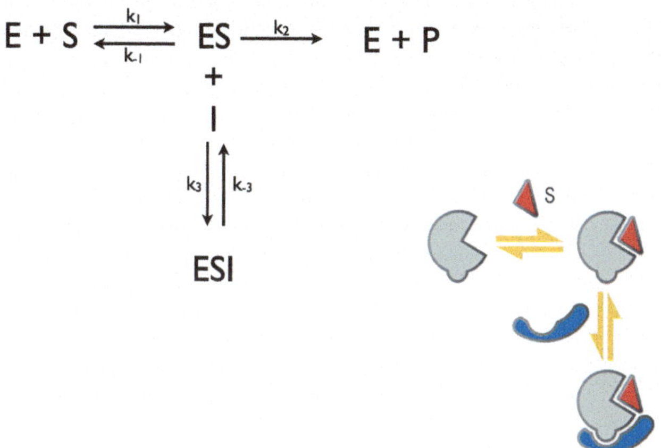

Fig. 7.11 Noncompetitive inhibition

7.3.2.2 Uncompetitive Inhibition

In this case, the inhibitor reacts reversibly only with the enzyme-substrate complex,[5] as shown in the kinetic scheme depicted in Fig. 7.11.

Under steady-state conditions ($\frac{dC_E}{dt} = \frac{dC_{ES}}{dt} = \frac{dC_{ESI}}{dt} = 0$), the following balances can be written:

$$C_E^0 = C_E + C_{EI} + C_{ES} \tag{7.79}$$

$$\frac{dC_E}{dt} = 0 = -k_1\, C_E\, C_S + k_{-1}\, C_{ES} + k_2\, C_{ES} \tag{7.80}$$

$$\frac{dC_{ES}}{dt} = 0 = k_1\, C_E\, C_S - k_{-1}\, C_{ES} - k_2\, C_{ES} - k_3\, C_{ES}\, C_I + k_{-3}\, C_{ESI} \tag{7.81}$$

$$\frac{dC_{ESI}}{dt} = k_3\, C_{ES}\, C_I - k_{-3}\, C_{EI} \tag{7.82}$$

From Eq. 7.82, it follows that $C_{ESI} = C_{ES}\,\frac{C_I}{K_I}$, where the dissociation constant of the enzyme-inhibitor complex is introduced as $K_I = \frac{k_{-3}}{k_3}$. From Eq. 7.80, it can be derived that $C_{ES} = C_E\,\frac{C_S}{K_M}$, where the Michaelis-Menten constant is defined as

[5] From a molecular perspective, this corresponds to a situation where the formation of the complex results in a protein conformation change, making an active site available to the inhibitor on the surface.

in the case of no inhibitor ($K_M = \frac{k_{-1}+k_2}{k_1}$). By substituting these expressions into Eq. 7.79, it is obtained:

$$C_E = \frac{C_E^0}{1 + \frac{C_S}{K_M} + \frac{C_I\, C_S}{K_I\, K_M}} \tag{7.83}$$

$$C_{ES} = \frac{C_E^0\, C_S}{K_M + C_S \cdot \left(1 + \frac{C_I}{K_I}\right)} \tag{7.84}$$

As in the previous paragraph, the (initial) rate of product formation is defined as:

$$v = \frac{dP}{dt} = k_2\, C_{ES} = k_2 \frac{C_E^0\, C_S}{K_M + C_S \cdot \left(1 + \frac{C_I}{K_I}\right)} = \frac{\frac{k_2\, C_E^0}{1+\frac{C_I}{K_I}} \cdot C_S}{\frac{K_M}{1+\frac{C_I}{K_I}} + C_S} = \frac{v'_{MAX}\, C_S}{K'_M + C_S} \tag{7.85}$$

where $v'_{MAX} = v_{MAX} \cdot \frac{1}{1+\frac{C_I}{K_I}}$ and $K'_M = K_M \cdot \frac{1}{1+\frac{C_I}{K_I}}$ represent the kinetic parameters modified by the presence of the inhibitor I compared to the kinetic parameters K_M and v_{MAX} of the enzyme kinetics in the absence of inhibition. It is worth noting that in this case, not only does the maximum rate of product formation decrease, but the constant K_M also decreases by the same amount: in this case, the slope of the line representing the velocity, which is linear at low substrate concentrations, remains unchanged. This means that when the substrate concentration is very low, the effects of the inhibitor's presence are not noticeable.

7.3.3 Noncompetitive Inhibition

In this case, the binding between the inhibitor and the enzyme occurs at a different site than the active site (effective for product formation). Unlike the case described in the previous paragraph, in this case, the inhibitor bound to the free enzyme prevents binding with the product, and similarly, the ES complex bound to the inhibitor I does not lead to product formation. The kinetic scheme is illustrated in Fig. 7.12.

As in the previous cases, under steady-state conditions, the following relationships hold:

$$C_E^0 = C_E + C_{EI} + C_{ES} + C_{ESI} \tag{7.86}$$

$$\frac{dC_E}{dt} = 0 = -k_1\, C_E\, C_S + k_{-1}\, C_{ES} + k_2\, C_{ES} \tag{7.87}$$

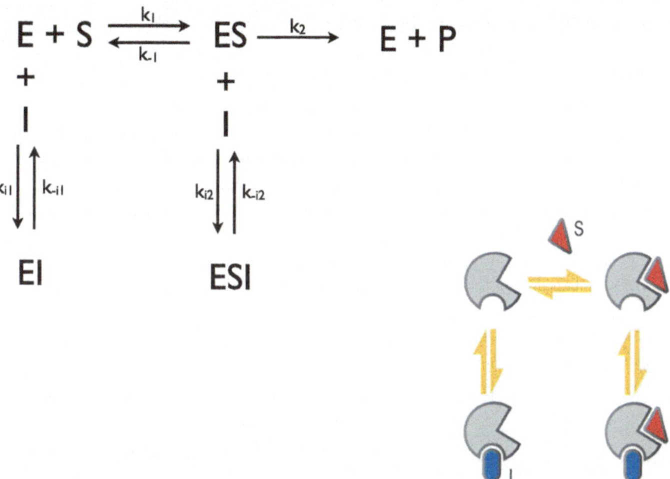

Fig. 7.12 Noncompetitive inhibition

$$\frac{dC_{ES}}{dt} = 0 = k_1\, C_E\, C_S - k_{-1}\, C_{ES} - k_2\, C_{ES} - k_{i2}\, C_{ES}\, C_I + k_{-i2}\, C_{ESI} \tag{7.88}$$

$$\frac{dC_{EI}}{dt} = 0 = k_{i1}\, C_E\, C_I - k_{-i1}\, C_{EI} \tag{7.89}$$

$$\frac{dC_{ESI}}{dt} = k_{i2}\, C_{ES}\, C_I - k_{-i2}\, C_{EI} \tag{7.90}$$

Introducing the two dissociation constants $K_I = \frac{k_{-i1}}{k_{i1}}$ and $K_I' = \frac{k_{-i2}}{k_{i2}}$, as well as the standard Michaelis-Menten constant $K_M = \frac{k_{-1}+k_2}{k_1}$, we find:

$$C_{EI} = \frac{C_E\, C_I}{K_I} \tag{7.91}$$

$$C_{ES} = \frac{C_E\, C_S}{K_M} \tag{7.92}$$

$$C_{ESI} = \frac{k_{-1} + k_2 + k_{i2}\, C_I}{k_{-i2}}\, C_{ES} - \frac{k_1}{k_{-i2}}\, C_E\, C_S = \frac{C_I\, C_E\, C_S}{K_M\, K_I'} \tag{7.93}$$

By substituting Eqs. 7.91–7.93 into Eq. 7.86, we obtain:

$$C_E = \frac{C_E^0}{\left(1 + \frac{C_I}{K_I}\right) + \frac{1}{K_M}\left(1 + \frac{C_I}{K_I'}\right) C_S} \tag{7.94}$$

Defining, as usual, the (initial) rate of product formation:

$$v = \frac{dP}{dt} = k_2 C_{ES} = k_2 \frac{C_E^0 C_S \cdot \left(1 + \frac{C_I}{K_I'}\right)}{K_M \cdot \frac{\left(1+\frac{C_I}{K_I}\right)}{\left(1+\frac{C_I}{K_I'}\right)} + C_S} = \frac{v_{MAX}' C_S}{K_M' + C_S} \tag{7.95}$$

where the modified kinetic parameters are defined as $K_M' = K_M \cdot \frac{\left(1+\frac{C_I}{K_I}\right)}{\left(1+\frac{C_I}{K_I'}\right)}$ and $v_{MAX}' = v_{MAX} \cdot \left(1 + \frac{C_I}{K_I'}\right)$, with K_M and v_{MAX} being the parameters of the Michaelis-Menten expression in the absence of inhibition. In the frequently occurring case[6] where $K_I \simeq K_I'$, it is also found that $K_M \simeq K_M'$.

The kinetic data related to enzymatic reactions can be evaluated in various ways, typically available in the form of the product formation rate, v, as a function of substrate concentration, C_S. The most commonly used representation is the *Lineweaver-Burk plot*:

$$\frac{1}{v} = \frac{K_M}{v_{MAX}} \cdot \frac{1}{C_S} + \frac{1}{v_{MAX}} \tag{7.96}$$

The deviation from the expected line in the absence of inhibition is shown in the figure for the three types of competition analyzed.

Additional Resources and Recommended Literature

The general definition of protein binding can be found in [3, 5]. The description of allostery and methods to its analysis is provided by [7, 33, 39].

[6] This case occurs when the site of interaction with the inhibitor is much different from the substrate-binding site on the enzyme surface, causing the dissociation constant of the complex to remain unchanged in the presence or absence of bound substrate.

Part II

Computational Biology

Molecular Dynamics Simulations for Computational Biology

8

You take the blue pill... the story ends, you wake up in your bed and believe whatever you want to believe. You take the red pill... you stay in Wonderland, and I show you how deep the rabbit hole goes – Morpheus (The Matrix)

L. & L. Wachowski

Abstract

Why it is important to know this material?

Protein binding is implied in all biological mechanisms, and the definition of quantitative models and laws is essential to assess in a quantitative way the protein binding to small ligands, proteins, and nucleic acids.

What is the key idea?

Starting from molecular models, it is possible to determine the quantitative laws for protein binding.

What is necessary to know already?

It is essential to know how the protein structure is built and how it interacts with the environment, through a delicate interplay between stability and adaptability.

8.1 Introduction

The tight link between protein structure and function is a central topic in protein science. The elective method to understand how the molecular structure instructs the functionality of proteins is to analyze the protein dynamics at the atomic level.

Molecular dynamics (MD) is a computational method that can simulate the motion of molecules over time, providing insights into their behavior at a level of detail that is difficult to achieve experimentally. In the context of proteins, MD

L. Di Paola, *Fundamentals of Molecular Bioengineering*,
https://doi.org/10.1007/978-3-031-42022-1_8

simulations can reveal how the three-dimensional structure of a protein changes over time, how it interacts with its environment, and how it undergoes conformation changes in response to various environment *stimuli*.

In this chapter, we will describe the principles and methods of molecular dynamics simulations for proteins. We will start by introducing the basic concepts of MD, including the force fields that define the interactions between atoms, the numerical algorithms used to integrate the equations of motion, and the statistical methods used to analyze the results. We will then discuss how these concepts are applied to model proteins, including the challenges posed by their size and complexity, as well as the strategies used to address these challenges.

Overall, this chapter aims to provide a comprehensive overview of the molecular dynamics of proteins, highlighting its strengths, limitations, and potential for advancing our understanding of protein structure and function. This chapter is aimed to provide the elementary elements of molecular dynamics (MD) for biomolecules with a perspective to applications in the protein science.

8.2 Basic Concepts in Molecular Dynamics

The molecular dynamics (MD) is the simulation of motion of atoms in molecules at an atomic level. Atoms in the molecule move according to interactions with other atoms and their trajectories are derived over a given lapse of time from the Newton's law, applied to the i-th atom in the system (being N the total number of atoms):

$$\mathbf{F}_i = m_i \cdot \mathbf{a}_i \tag{8.1}$$

the force $\mathbf{F}_i$ is due to the interaction with other atoms. In this sense, molecular dynamics is a *deterministic* method, since the trajectories of atoms are exactly determined by the integration of the classical dynamics laws. The atoms trajectories are traced in the 6N-dimensional space ($3N$ positions and $3N$ momenta). According to this information, it is possible to write for the molecule the *Hamiltonian* $H(\Gamma)$, being Γ the whole set of positions and momenta for all atoms.

Actually, in protein molecular dynamics, atom trajectories are not particularly informative as the protein molecule moves between different conformations. Instead, it is more compelling to gather all possible conformations based on specific constraints, such as maintaining a constant total energy of the system E. This approach shifts the focus to a *statistical mechanics perspective*, where the emphasis is on collecting data on the conformation ensemble, as aptly and fully developed in the following subsection.

Finally, it is important to stress out that, despite being an effective technique, in some cases the only available to gather information about protein dynamics, molecular dynamics has some limitations, such as:

- *Use of classical forces*: at atom level, the interactions between atoms are driven by quantum forces rather by classical Newton forces, which are, conversely, a good approximation at temperatures close to room temperature while fail to describe systems at very low temperature; the temperature threshold depends also on molecular mass, so for small molecules, the description through classical forces is poor, while the description of large biomolecules at room temperature with classical forces is a good approximation;

- *Realistic interaction forces*: the interaction forces between atoms are determined as gradient of an *interaction potential*, so the reliability of forces depends on the choices of the interaction potential, which should describe as better as possible the real interactions between atoms while keeping the computational burden at a reasonable level;
- *Time and size constraints*: a good description of the molecular system should catch the order of magnitude of times and size of the simulation box; if the elapsed time of the simulation is chosen lower than the typical relaxation times of the real molecular system, the MD simulation is likely to catch only random oscillations losing the perspective of the whole process duration; in the same way, a wrong choice of the size of the observation box can mask crucial phenomena affecting the molecular system dynamics (such as the effect of long-range electrostatic potentials). In this case, the reference is given by the correlation lengths of the correlation functions of interest in the space.

All these limitations are the typical limitations of a physical model, whose efficacy in describing the real physical system depends on the underlying approximations. When designing a model to study a complex process, it is important to strike a balance between accuracy and computational efficiency. If the model is too restrictive, it may fail to capture important aspects of the process, leading to poor results. On the other hand, a model with a very loose approximation can lead to a significant increase in computational burden, resulting in prohibitively long simulation times to obtain results.

In practice, the level of approximation must be chosen carefully based on the specific system being studied and the research questions being asked. Generally, models that are too restrictive should be avoided, as they are unlikely to provide useful insights. However, models with a very loose approximation can be useful if the computational resources are available and if the added accuracy is necessary to answer specific research questions.

Ultimately, the goal of any model, including that behind molecular dynamics (MD) simulations, is to provide a useful and accurate representation of the system being studied. Striking the right balance between approximation and computational efficiency is crucial in achieving this goal and obtaining meaningful results from simulations.

8.2.1 Design Constraints and Ensembles

In molecular dynamics, design constraints play a crucial role in aligning computational power with the accurate representation of real-world processes. Two key factors that must be taken into account are time and space. In the case of molecular dynamics simulations, it is essential to consider the *timescale of the process* being studied, as a successful simulation should match this timescale. For instance, the chemical reactions for small organic molecules typically occur in picoseconds to nanoseconds, while molecular processes, such as protein folding, can occur even in milliseconds, typically within nanoseconds and microseconds. For large biomolecules, such as proteins, the current CPU power allows for all-atom simulations lasting for weeks, describing the dynamic evolution of molecular systems over microseconds. Another crucial point is to choose *the right timestep*, i.e., the sampling interval, to accurately describe the kinetic process. In molecular dynamics simulations, the integration timestep refers to the interval at which the equations of motion are numerically integrated. So the whole simulation time should cover all the process, but the timestep should be proper to describe the variations of key properties leading the time evolution of the system. MD simulation timestep is a critical parameter that affects the accuracy, stability, and computational efficiency of the simulation.

In a classical MD simulation, the most computationally intensive task is the evaluation of the *potential energy* as a function of Γ with the nonbonded or non-covalent interactions being the most expensive part. When all pair-wise electrostatic and van der Waals interactions are accounted for explicitly, the computational cost scales as $O(n^2)$, where n is the number of particles. However, this cost can be reduced by using methods such as *particle mesh Ewald summation* ($O(nlog(n))$), *particle-particle-particle-mesh* (P3M), or *good spherical cutoff methods* ($O(n)$). These electrostatic methods help to reduce the computational cost associated with MD simulations.

During each timestep, the forces acting on each particle are calculated based on the current positions and velocities of the particles. These forces are then used to update the positions and velocities of the particles using an integration scheme, such as *Verlet* or *leapfrog integration*. The accuracy of this numerical integration scheme depends on the timestep size, with smaller timesteps generally leading to more accurate results but higher computational costs.

The timestep size is also closely related to the potential energy evaluation in molecular dynamics simulations. Specifically, the forces acting on each particle are calculated based on the gradient of the potential energy with respect to the particle's position. Therefore, when using a shorter timestep, more frequent evaluations of the potential energy are required to accurately capture the dynamics of the system. Conversely, when using a longer timestep, fewer evaluations of the potential energy are needed, but the accuracy of the simulation may suffer. Finding the optimal timestep size is therefore an important consideration in molecular dynamics simulations. Similarly, the choice of *system size* in molecular dynamics simulations has a significant impact on both the accuracy of the simulation and the associated computational cost. The system size is typically defined by the number of particles in the simulation and the dimensions of the simulation box, which can be *periodic* or *non-periodic*.

When a periodic boundary condition is used, the simulation box is replicated in all directions, resulting in an infinite lattice of identical boxes. This allows for the simulation of an infinitely large system while keeping the computational cost reasonable while avoiding also boundary condition artifacts.

In contrast, nonperiodic boundary conditions can be used to simulate a finite system with fixed boundaries, which can be more suitable for some types of systems, such as those with a well-defined surface (materials or cell membranes). However, simulations with nonperiodic boundary conditions are generally more computationally expensive, since a larger system size is required to reduce the influence of boundary effects.

Usually, protein MD simulations are carried out using periodic boundary conditions, where the cell unit represents the solvated protein molecule (protein molecule plus solvation molecules).

The size of the system also affects the computational cost of the simulation, since a larger system requires more computational resources to simulate. However, the size of the system must be chosen carefully to balance the need for accuracy with the available computational resources. In practice, the size of the system is often determined by considering the system's properties and the specific research question being investigated.

Eventually, protein molecular dynamics requires to account for solvent; as mentioned in the following chapter, protein solvation is a key factor for protein stability in physiological environment, and therefore the solvent is expected to play a central role in protein dynamics (Fig. 8.1 represents the box of a protein (Src kinase) surrounded by water molecules)

To simulate the effect of solvent molecules *explicit* or *implicit*, solvent models can be used, where the explicit choice requires a much larger computational burden to explicitly account for the interactions between solvent molecules and

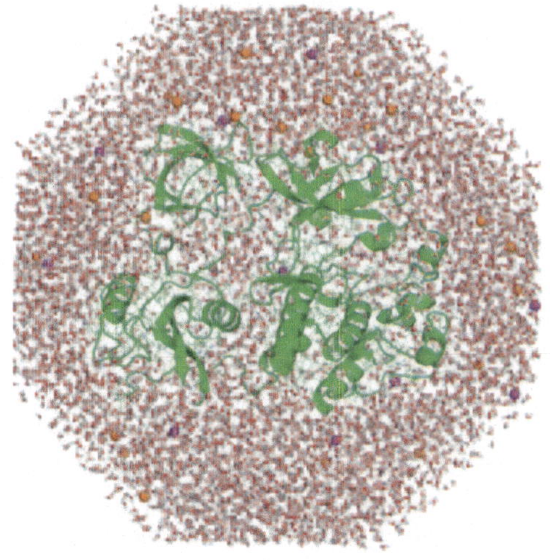

Fig. 8.1 Simulation of a solvated protein: Src kinase protein (green) immersed in ~15000 water molecules (oxygen in red, hydrogen in white). Reprinted with permission from [28]

the biomolecule (protein). On the other hand, the implicit solvent model is equivalent to the abovementioned *mean field approximation* when globally water molecules constitute a mean homogeneous field where the protein molecules can interact to each other and the protein interactions is modulated by the presence of water molecules as mean environment. This option is much lighter in terms of computational times, but the granularity of the solvent is strictly required in many applications that involve studying the detailed interactions between a protein and its surrounding solvent molecules. In particular, explicit solvent modeling is essential for accurately predicting the thermodynamics and kinetics of protein folding, as well as for studying the binding of ligands to proteins.

One specific example where the *granularity of the solvent* is required is in the study of protein-ligand binding. In this case, explicit solvent modeling allows for the accurate representation of the solvation effects that can influence the binding affinity between the protein and the ligand. By including explicit water molecules in the simulation, the solvation effects on the protein and the ligand can be explicitly accounted for, which is crucial for predicting the binding affinity with high accuracy.

Another example where the granularity of the solvent is required is in the study of protein-protein interactions. In this case, the explicit inclusion of solvent molecules can provide insights into the detailed interactions between the two proteins and how the solvent affects the interaction. In particular, solvent molecules can mediate interactions between the two proteins and can play a role in stabilizing the protein-protein complex.

All in all, the critical choices dictated by the design constraints influence the adherence of the simulated systems with the real one, given also the limitations of the available computational resources.

Despite these design constraints and computation limitations, molecular dynamics simulations have become a benchmark to gain insight into the behavior of proteins and their interactions with other molecules. With continued advancements in computational power and simulation algorithms, molecular dynamics simulations will continue to play an important role in the study of protein structure and function. In the following, some different types of ensemble used in MD simulations will be shown in details to better frame the commonest applications, from biotechnology to drug discovery.

Finally, the choice of ensemble in molecular dynamics (MD) simulations of proteins is a critical step having a significant impact on the behavior and properties of the simulated system. The ensemble determines the type of thermodynamic variables that are controlled during the simulation, such as temperature, pressure, and particle number. Here are some key points highlighting the importance of the *ensemble choice*:

- *System Representation*: The ensemble choice affects how the protein system is represented and the conditions under which it is simulated. The NVT ensemble, as shown later, keeps the temperature constant, which is important for studying temperature-dependent properties such as protein folding, stability, and conformation transitions. The NPT ensemble, on the other hand, introduces additional variables such as pressure and volume, which are essential for simulating proteins in a solvent environment and capturing their behavior under realistic conditions.

- *Physical Realism*: The ensemble choice should reflect the physical conditions that are relevant to the system under study. Proteins typically exist in an aqueous environment, and simulating them under constant pressure (NPT ensemble) allows for a more realistic representation of their behavior in solution. This is particularly important when studying phenomena such as protein-protein interactions, ligand binding, or protein dynamics in membrane environments.
- *Equilibrium Sampling*: The choice of ensemble influences the type and extent of equilibrium sampling that can be achieved. Different ensembles provide different types of equilibrium and allow exploration of different regions of the system's phase space. The NVT ensemble, for instance, is particularly suited for sampling the conformation landscape of a protein at a specific temperature, which is essential for understanding its thermodynamics and kinetics.
- *Accuracy of Results*: Strictly tied to physical realism, the choice of ensemble affects the accuracy of computed properties and observables. Simulating proteins in an ensemble that closely resembles the experimental conditions (e.g., NPT ensemble) improves the accuracy of computed properties, such as protein-ligand binding affinities, solvation free energies, or protein conformation ensembles. Using an appropriate ensemble can also help reproduce experimental observables, like protein structural fluctuations or thermodynamic properties.
- *Computational Efficiency*: The ensemble choice can also influence the computational efficiency of the simulations. Some ensembles, such as the NPT ensemble, require additional algorithms (e.g., pressure control algorithms) that introduce extra computational cost. Depending on the research question and available computational resources, choosing an ensemble that balances accuracy and efficiency is crucial.

In summary, the choice of ensemble in MD simulations of proteins plays a critical role in determining the level of realism, equilibrium sampling, accuracy of results, and computational efficiency. It is important to carefully consider the objectives of the study, the properties being investigated, and the physical conditions under which the protein operates when selecting the appropriate ensemble. In the following, the commonest ensembles used in molecular dynamics of proteins are described, with some key applications in the protein science field.

8.2.1.1 Microcanonical and Canonical Ensemble

As stated in Sect. 2.4, the **microcanonical ensemble** represents protein conformation states whose energy (E), number of molecules (N), and volume (V) are fixed, corresponding to thermodynamic *isolated* systems. Any transformation in the microcanonical ensembles is *adiabatic* (no heat exchange) and *isochoric* (constant volume), and from a microscopic point of view, it can be described as an exchange between kinetic and potential energy, being the total mechanical energy E constant. The atoms trajectory during transformation is described in terms of $\mathbf{X}(t)$ and $\mathbf{V}(t)$, respectively, the atom's coordinates and velocity, derived from the integration of the

Newton's laws for motion:

$$
\begin{aligned}
F(\mathbf{X}) &= -\nabla W(\mathbf{X}) = M \cdot \dot{\mathbf{V}} \\
\mathbf{V}(t) &= \dot{\mathbf{X}(t)}
\end{aligned}
\tag{8.2}
$$

where $W(\mathbf{X})$ is the *potential energy function* (or, simply, *potential*), which depends on the atoms coordinates X; the first equation is the Newton's law of motion, and the second is the definition of velocity in terms of time derivative of spatial coordinates. The force acting on any molecule's atom F depends on atom's spatial coordinates $\mathbf{X}$ and can be computed as negative gradient of the potential energy function $W(\mathbf{X})$. Equation 8.2 must be integrated through a numerical algorithm to provide as results numerical value of positions $\mathbf{X}(t)$ and velocities $\mathbf{V}(t)$ for each atom (atom's trajectory).

The numerical integration of Eq. 8.2 must preserve the physical representation of the system, looking only for reasonable, realistic solutions. Purposed algorithms are used, which are adapted to the solution of dynamical problems for Hamiltonian systems, called *symplectic integrators*.

Among these, the widely used *Verlet integration method* is valued for its simplicity, computational efficiency, and ability to accurately preserve energy conservation and phase space volume, making it suitable for modeling conservative systems like proteins. The Verlet integration algorithm is a second-order symplectic integrator that updates the positions and velocities of particles in discrete timesteps. It is based on Taylor series expansions and approximations of the equations of motion. The key idea behind the Verlet method is that it utilizes information from previous time steps to compute the positions and velocities at the current time step.

The Verlet integration method has several advantages that make it well-suited for protein MD simulations:

- *Time Reversible and Symplectic*: The Verlet method is time-reversible, meaning that the dynamics can be accurately reversed by simply negating the velocities. This property allows for the exploration of both forward and backward trajectories, enabling calculations involving time-reversal symmetry. Additionally, the Verlet method is symplectic, preserving the system's symplectic structure[1] and energy conservation.
- *Conserves Energy*: One of the crucial properties of the Verlet method is its excellent energy conservation. It accurately maintains the total energy of the

[1] The symplectic space refers to a vector space equipped with a symplectic form, which is a nondegenerate, closed, and skew-symmetric bilinear form. The symplectic form captures the geometric and algebraic structure of the space, providing a way to measure areas in phase space.

In simpler terms, a symplectic space is a mathematical framework used to study systems with continuous variables, such as position and momentum in physics. It allows for the analysis of the dynamics, stability, and conservation laws in these systems. Symplectic spaces have applications in various fields, including physics, mathematics, and computational biology, where they play a crucial role in understanding and modeling complex systems.

system during the simulation, which is essential for capturing the dynamics and equilibration of proteins accurately, with a special regard for the microcanonical ensemble choice.
- *Numerical Stability*: The Verlet method exhibits good long-term numerical stability, meaning that it can reliably maintain the accuracy of simulations over extended periods. The algorithm's stability helps prevent the accumulation of errors, enabling longer MD simulations and improved exploration of the conformation space.
- *Computational Efficiency*: The Verlet integration algorithm requires minimal computational resources and is computationally efficient compared to higher-order integration methods. Its simplicity allows for fast calculations and facilitates the simulation of large-scale systems or long-time trajectories.

Despite its advantages, it's important to note that the Verlet method is not suitable for all types of simulations. It assumes that the forces acting on the particles are time-independent within a given time step, making it less appropriate for systems with rapidly varying forces or nonconservative forces, such as those involving thermostats or external driving forces.

In the **canonical ensemble**, the corresponding thermodynamic systems have fixed number of molecules (N), volume (V), and temperature (T); for this reason, the canonical ensemble is also referred to as NVT.

In these systems, the energy exchange occurs with a perfect balance between exothermic and endothermic processes, so the temperature does not vary. So, the system is imagined as placed in a thermostat with a fixed temperature. The crucial step in the canonical ensemble description through MD is to define a *thermostat algorithm*, such as:

- *Langevin Dynamics*: The Langevin thermostat introduces a friction term and random forces to mimic the interaction between the protein and its surrounding solvent. This algorithm maintains the system at a desired temperature by balancing the friction and random forces.
- *Nosé-Hoover Chain*: The Nosé-Hoover thermostat is a deterministic algorithm that extends the equations of motion to include a set of auxiliary variables to improve its performance and accuracy.
- *Berendsen Thermostat*: The Berendsen thermostat is a simple and widely used algorithm that rescales the velocities of particles to control the temperature. It is computationally efficient but does not accurately represent the true dynamics of the system.
- *Andersen Thermostat*: The Andersen thermostat introduces stochastic collisions between the particles and a heat bath. It randomly selects a particle and replaces its velocity with a random value drawn from the Maxwell-Boltzmann distribution at the desired temperature.
- *Martyna-Tuckerman-Tobias-Klein (MTTK) Thermostat*: The MTTK thermostat is an extension of the Nosé-Hoover chain algorithm. It employs multiple Nosé-Hoover chains to achieve improved energy conservation and ergodicity.

- *Velocity Rescaling*: Velocity rescaling is a simple thermostat algorithm where the velocities of particles are scaled at each timestep to match the desired temperature. However, this algorithm does not account for energy exchange with the surroundings.

The reliability of these algorithms is fairly good in providing realistic results; however, the research is strongly working on improving their reliability and the computational efficiency.

Here are some situations where adopting a canonical ensemble is appropriate and convenient:

- *Equilibrium Sampling*: When studying the equilibrium properties of a protein system, such as conformation dynamics, thermodynamic properties, or ligand binding, the canonical ensemble is often used. It allows the system to explore different energy states while maintaining a constant temperature.
- *Temperature-dependent Studies*: If you are interested in investigating the temperature dependence of protein behavior or properties, the canonical ensemble is suitable. By performing simulations at different temperatures while keeping the number of particles and volume constant, you can study the effects of temperature on protein dynamics, stability, folding, and other phenomena.
- *Comparisons and Consistency*: The canonical ensemble provides a common framework for comparing different systems or different stages of a simulation. By using the same temperature for multiple simulations, you can make meaningful comparisons between different protein structures, mutants, or ligand binding states.
- *Replica Exchange Molecular Dynamics (REMD)*: REMD is a powerful sampling technique that combines multiple simulations at different temperatures. Each replica in REMD operates in the canonical ensemble, allowing for enhanced sampling of conformation space and exploration of rare events. REMD is particularly useful for studying protein folding, protein-protein interactions, or exploring energy landscapes.
- *Thermodynamic Analysis*: When analyzing the thermodynamics of a protein system, such as calculating free energy differences, using the canonical ensemble is essential. By applying techniques like umbrella sampling, thermodynamic integration, or free energy perturbation, the canonical ensemble is required to obtain accurate thermodynamic quantities.

On the other hand, as follows are listed some classes of application where using the microcanonical ensemble is suitable:

- *Energy Conservation*: The microcanonical ensemble is ideal when studying energy-conserving systems, where the total energy remains constant. This ensemble is useful for investigating energy transfer, energy landscapes, or exploring the dynamics of isolated proteins.

- *Transition State Analysis*: When examining reaction pathways, protein folding/unfolding, or conformation transitions, the microcanonical ensemble can be useful to explore the potential energy surfaces, identify transition states, and characterize reaction kinetics.
- *Vibrational Analysis*: The microcanonical ensemble is suitable for studying the vibrational properties of proteins, including normal mode analysis or calculation of vibrational spectra. It enables the investigation of protein motions and collective vibrations without temperature fluctuations.
- *Energy-Dependent Phenomena*: If the properties of the system are primarily governed by energy-dependent processes, such as thermally activated reactions or energy barriers, the microcanonical ensemble can provide insights into these phenomena.

It's important to highlight that the microcanonical ensemble is not as commonly used as the canonical ensemble in MD simulations of proteins, as biological systems are often in contact with a heat bath (e.g., water) and temperature control is crucial. However, in certain specific cases where energy conservation or energy-dependent phenomena are of interest, the microcanonical ensemble can be employed effectively.

At a glance, it is evident the choice between the microcanonical and canonical ensemble is crucial not only to have a reliable and realistic representation of the physical systems but also to catch the relevant dynamical process in an appropriate fashion.

8.2.1.2 Isothermal-Isobaric (NPT) and Generalized Ensembles

The microcanonical and canonical ensembles cover the majority of applications of MD simulations for protein systems. However, in the field of molecular dynamics (MD) simulations of proteins, the isothermal-isobaric ensemble, also known as the NPT ensemble, and the generalized ensemble play important roles in studying protein systems under constant temperature, pressure, and other thermodynamic conditions.

The isothermal-isobaric ensemble (NPT) represents thermodynamic systems where are fixed the number of molecules (N), pressure (P), and temperature (T), representing the laboratory conditions of a flask exposed to air at room temperature and pressure. In contrast to the canonical ensemble, the computation in the isothermal-isobaric (NPT) ensemble necessitates the incorporation of a barostat. The role of the barostat is to regulate the influence of pressure on the system by modifying the volume, accomplished through a minor rescaling of the atom positions.

Several barostat algorithms are commonly used for maintaining constant pressure in NPT ensemble molecular dynamics (MD) simulations of proteins. Here are some of the main barostats:

- *Berendsen Barostat*: The Berendsen barostat is similar in concept to the Berendsen thermostat. It is a simple and widely used algorithm that couples the system to a pressure bath to maintain a desired pressure. It adjusts the system volume according to the pressure coupling strength but does not accurately reproduce the equilibrium distribution of states.

- *Parrinello-Rahman Barostat*: The Parrinello-Rahman barostat is a more sophisticated algorithm that dynamically scales the system volume while maintaining constant pressure. It provides better equilibration and sampling of the phase space.
- *Martyna-Tobias-Klein Barostat*: The Martyna-Tobias-Klein (MTK) barostat is an extension of the Parrinello-Rahman algorithm that improves its performance and robustness. It uses a chain of oscillators to control the volume fluctuations and better conserve energy.
- *Monte Carlo Barostat*: The Monte Carlo barostat employs a probabilistic approach to adjust the system volume and pressure. It uses trial volume changes and Metropolis acceptance/rejection criteria to equilibrate the system at the desired pressure.

These are some of the main barostat algorithms used in NPT ensemble MD simulations of proteins. The choice of barostat depends on factors such as the desired accuracy, computational efficiency, and the specific requirements of the protein system being studied.

The NPT ensemble is particularly useful for studying protein conformation changes, protein-ligand interactions, and the behavior of proteins in solution or under external stress conditions.

8.2.1.3 Generalized Ensemble in MD Simulations

Generalized ensemble in molecular dynamics (MD) of proteins is a set of computational methods that are used to overcome the limitations of traditional MD simulations. Traditional MD simulations are limited by the fact that they can only sample a small portion of the conformation space of a protein. This is because the free energy landscape of a protein is very rugged, with many local minima. Generalized ensemble methods allow for more efficient sampling of the conformation space by introducing bias into the simulation. This bias can be in the form of temperature, pressure, or other thermodynamic variables.

There are many different generalized ensemble methods, but some of the most common include the following:

- Replica exchange MD (REMD);
- Parallel tempering MD (PTMD);
- Multicanonical MD (MCMD);
- Wang-Landau sampling;
- Adaptive biasing force (ABF)

Generalized ensemble methods have been used to study a wide variety of protein properties, including folding, binding, and catalysis. They have also been used to study the effects of mutations on protein structure and function.

Here are some of the advantages of using generalized ensemble methods in MD simulations of proteins:

- They can sample a larger portion of the conformation space, which can lead to more accurate results.
- They can be used to study systems that are difficult to study with traditional MD simulations, such as large proteins and proteins with multiple states.
- They can be used to study systems that are far from equilibrium, such as proteins that are folding or unfolding.

Here are some of the disadvantages of using generalized ensemble methods in MD simulations of proteins:

- They can be computationally expensive.
- They can be difficult to implement.
- They can introduce bias into the simulation, which can lead to inaccurate results.

Overall, generalized ensemble methods are a powerful tool for studying the structure and function of proteins. They can be used to study systems that are difficult or impossible to study with traditional MD simulations. However, it is important to be aware of the limitations of these methods before using them. Lu

8.2.2 Molecular Simulations Force Fields and Potentials

The definition of the interaction potentials and force fields plays a central role for molecular dynamics (MD) simulations of proteins. They provide a way to calculate the forces between atoms, which are necessary to determine the motion of the atoms over time. There are many different interaction potentials and force fields available, and the choice of which one to use depends on the specific system being simulated.

Some of the factors that need to be considered when choosing an interaction potential or force field include the following:

- The size of the system being simulated.
- The level of accuracy required.
- The computational resources available.

For small systems, such as small molecules, it is often possible to use a simple interaction potential, such as a Lennard-Jones potential. However, for larger systems, such as proteins, more complex interaction potentials are needed. These more complex interaction potentials can take into account the effects of charge, polarizability, and dispersion.

As for the accuracy, for some applications, such as studying the folding of a protein, it is important to use an interaction potential or force field that is very accurate. However, for other applications, such as studying the dynamics of a protein in solution, it may be sufficient to use an interaction potential or force field that is less accurate.

The computational resources available also need to be considered when choosing an interaction potential or force field. The choice of some interaction potentials and force fields is more computationally expensive than others. It is crucial to find a trade-off between the accuracy and the computational costs when choosing an interaction potential or force field that is affordable in terms of the computational resources available.

Here are some additional points to consider when choosing an interaction potential or force field for MD simulations of proteins, similar to general consideration when choosing the most appropriate ensemble, for instance:

- The interaction potential or force field should be based on a sound theoretical foundation.
- The interaction potential or force field should be consistent with experimental data.
- The interaction potential or force field should be computationally efficient.

By considering all of these factors, it is possible to choose an interaction potential or force field that is well-suited for MD simulations of proteins.

8.2.2.1 Empirical and Semiempirical Potentials

In molecular dynamics (MD) simulations of proteins, the choice of potential energy functions is crucial for accurately describing the behavior and interactions of protein molecules.

Empirical potentials and semiempirical potentials are two types of mathematical models used in MD to approximate the potential energy of protein systems.

Empirical potentials, commonly known as *force fields*, are mathematical functions used to approximate the potential energy of a protein system based on empirical data and observations. These potentials rely on parameters derived from experimental data and quantum mechanical calculations. They are designed to reproduce various properties of proteins, including bond lengths, bond angles, torsion angles, and non-covalent interactions.

Empirical potentials typically consist of terms representing bonded interactions (bond stretching, angle bending, and dihedral rotations) as well as nonbonded interactions (van der Waals forces and electrostatic interactions). These potentials are derived from a combination of experimental measurements, quantum mechanical calculations, and fitting to reproduce the observed behavior of proteins.

The parameters in empirical potentials are often generalized to be transferable across different protein systems, allowing for simulations of diverse proteins. Popular force fields used in protein MD simulations include AMBER, CHARMM, and OPLS.

On the other hand, **semiempirical potentials** combine elements of both empirical force fields and quantum mechanical calculations. These potentials are particularly useful for studying specific aspects of proteins where quantum mechanical effects play a significant role.

In the context of protein MD simulations, semiempirical potentials are often employed in the form of quantum mechanics/molecular mechanics (QM/MM) approaches. QM/MM combines a quantum mechanical treatment of a small region (e.g., active site, metal ion, covalent bonds) using methods like *density functional theory* (DFT) or calculations with an empirical force field treatment of the rest of the protein system. This allows for the accurate description of electronic structure, charge transfer, and chemical reactions occurring in the localized region of interest.

Semiempirical potentials require more computational resources compared to purely empirical force fields due to the inclusion of quantum mechanical calculations. They are particularly valuable for studying enzymatic reactions, active sites, metalloproteins, and other systems involving specific quantum mechanical effects.

Overall, both empirical and semiempirical potentials play essential roles in protein MD simulations, with empirical potentials providing a balance between accuracy and computational efficiency, while semiempirical potentials offer more detailed quantum mechanical descriptions for specific regions of interest.

8.2.3 Coarse-Grained Molecular Dynamics of Proteins

The all-atom MD simulations of protein molecules provide a very complete and comprehensive description of the protein dynamics and help predicting protein folding in an exhaustive way for proteins of unknown structure. Unfortunately, the computational complexity grows with $O(n^2)$, where n is the number of atoms, due to the pairwise interactions between atoms, which must be accounted for. On the other hand, slow processes involving large molecules or aggregates are managed over very long times, paving the way to a new approach to molecular dynamics, where the loose of fine details is the price to pay for a broader and long-term perspective.

Figure 8.2 represents the scale of size and time for protein processes. It is evident that the protein processes involving larger molecular structures (multichain proteins or aggregates) or slower processes (such as aggregation) require enormous computation times and so the choice to reduce the level of details to have a reasonable picture in a reasonable time is often an obliged trade-off. The modeling approach used for protein systems, known as *CG* (CG) modeling, assumes that the primary events in protein dynamics can be represented by the overall movement of individual residues, specifically their heavy atoms (e.g., C atoms). In this approach, the detailed motion of side chains is not explicitly described but is influenced by the overall movement of the corresponding residue. Indeed, the *CG* approach of the protein MD is part of its very history.

The first CG model of protein dynamics is known as the *Gō model* or *Gō-like model*. It was introduced by H. A. Scheraga and coworkers in the late 1970s. The Gō model simplified the protein representation by considering a reduced set of

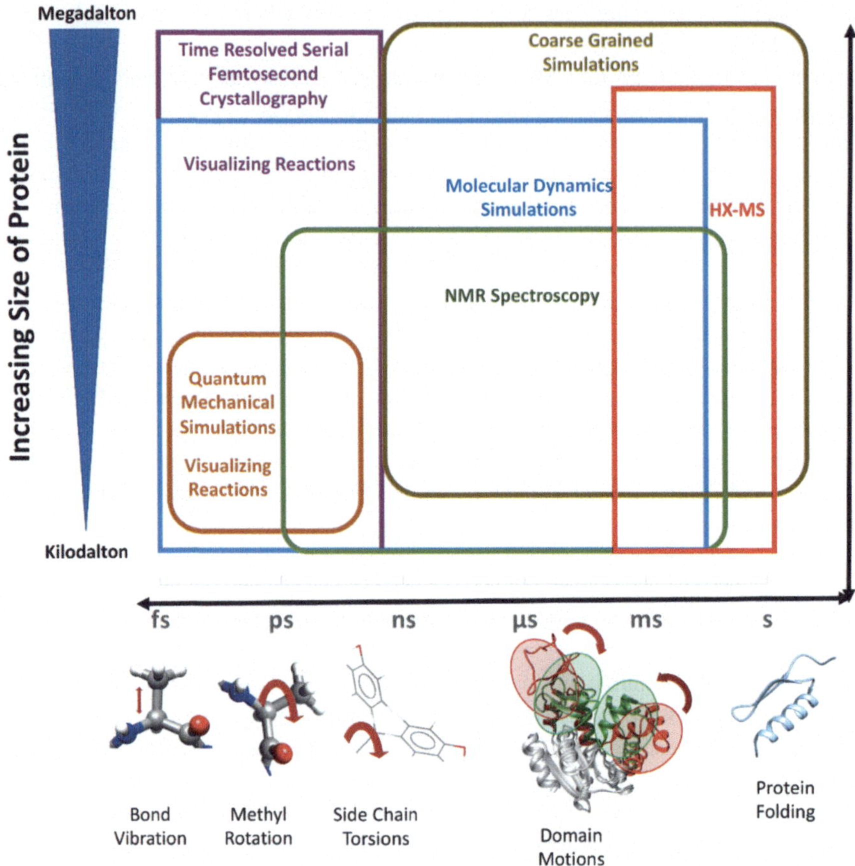

Fig. 8.2 Protein processes can occur over a wide range of time and size scales. Slower processes, like protein aggregation, can span months or even years and involve larger molecular structures. Describing these processes computationally poses a significant challenge as the computational burden increases steeply, often to a point that is practically unattainable. Reprinted with permission from [37]

interactions and consequently representing the protein structure in a CG, simplified way.

In the Gō model, each amino acid residue is represented by a single interaction center, typically the C atom. The interactions in the model are designed to capture the essential features of protein folding, focusing on the hydrophobic collapse and formation of native contacts. The model assumes a simplified energy function that favors native contacts and disfavors non-native contacts. This CG approach has been valuable for studying protein folding mechanisms and characterizing the thermodynamics of protein structures.

Since the development of the Gō model, various other CG models have been proposed, each with its own level of simplification and specific goals. These models range from simple elastic network models to more complex residue-based models with additional parameters capturing specific interactions or structural features.

It's important to note that while the Gō model was one of the earliest CG models for protein dynamics, subsequent advancements and refinements have led to the development of numerous other CG models that better capture the diverse aspects of protein behavior and dynamics.

Later, in 1976, Martin Karplus and Michael Levitt introduced a simplified approach to model protein dynamics using a CG representation.

The Karplus-Levitt model aimed to capture the collective motions and fluctuations of proteins while reducing the computational complexity. The model treated the protein as a set of rigid bodies connected by flexible torsional angles, neglecting the detailed atomistic interactions. The primary focus was on describing the low-frequency motions and capturing the essential dynamics relevant to protein function.

The model utilized a simplified potential energy function that incorporated torsional angle potentials to represent the bond rotations and elastic potentials to capture the inter-residue interactions. By incorporating these simplified potentials, Karplus and Levitt were able to simulate protein motions on a reasonable timescale using the computational resources available at the time.

The Karplus-Levitt model had a significant impact on the field of protein dynamics and helped pave the way for subsequent developments in the field. It provided insights into the collective motions of proteins and the relationship between structure and dynamics. Although the model has limitations due to its CG nature, it served as an important foundation for further advancements in understanding protein dynamics and conformation changes.

Since the introduction of the Karplus-Levitt model, more sophisticated and detailed approaches, such as atomistic molecular dynamics simulations, have been developed. However, the KL model played a crucial role in inspiring subsequent research and establishing the importance of protein dynamics in understanding biological function.

CG (CG) models continue to be highly relevant in molecular dynamics (MD) simulations due to their ability to capture the essential features of biomolecular systems while reducing computational complexity. Here are some key aspects that highlight the relevance of CG models in MD:

- *Increased Time and Length Scales*: CG models allow for the simulation of biomolecular systems at longer time and larger length scales compared to all-atom models. By reducing the number of particles and interactions, CG models enable the exploration of biologically relevant timescales, such as protein folding, conformation transitions, and assembly processes, that are difficult to reach with atomistic models, such as protein aggregation.
- *Efficient Sampling*: CG models provide a more efficient sampling of conformation space compared to all-atom models. By averaging out the fine details of atomic interactions, CG models allow for faster exploration of relevant states and

capture the dominant motions and interactions governing the system's behavior. This is particularly advantageous for studying large macromolecular assemblies or long-timescale processes.

- *Insight into Collective Dynamics*: CG models can capture the collective motions and large-scale conformation changes in biomolecules, shedding light on the dynamics and functional mechanisms that are challenging to observe at the atomic level. By focusing on the essential features and long-range interactions, CG models provide a simplified yet meaningful representation of biomolecular systems.
- *Parameterization and Model Development*: CG models offer flexibility in parameterization and model development, allowing for the customization of the model to specific biomolecular systems or research questions. CG models can be tailored to reproduce experimental data or target specific properties, making them versatile tools for studying various biological processes.
- *Integration with Experiments*: CG models can be used in conjunction with experimental techniques to interpret and complement experimental data. By simulating the behavior of biomolecular systems, CG models can provide insights into the dynamics, energetics, and mechanisms underlying experimental observations, aiding in the interpretation and understanding of experimental results.
- *Computational Efficiency*: CG models are computationally more efficient compared to all-atom models, as they involve fewer particles and interactions. This efficiency enables the simulation of larger systems and longer timescales within available computational resources.

Coarse-grained (CG) models in molecular dynamics (MD) have found numerous applications across a wide range of biological systems and processes. Here are some concrete applications of CG models in MD:

- *Protein Folding and Unfolding*: CG models are widely used to study the folding and unfolding pathways of proteins. By simplifying the representation of amino acids and focusing on the dominant interactions, CG models enable the simulation of larger proteins over longer timescales, providing insights into the thermodynamics and kinetics of protein folding.
- *Membrane Proteins and Lipid Bilayers*: CG models are employed to investigate the structure, dynamics, and interactions of membrane proteins in lipid bilayers. By coarse-graining lipids and proteins, these models enable simulations of membrane systems for extended timescales, allowing the exploration of membrane protein function, membrane-protein interactions, and membrane dynamics.
- *Protein-Protein Interactions*: CG models are used to study protein-protein interactions and the assembly of multi-protein complexes. Coarse-grained representations of proteins can capture the essential features and dynamics involved in complex formation, allowing for the exploration of binding pathways, conformation changes, and protein complex stability.
- *Nucleic Acids*: CG models are employed to investigate the structure, dynamics, and interactions of nucleic acids, such as DNA and RNA. These models can

capture the essential features of nucleic acids while reducing computational complexity, enabling the simulation of large DNA/RNA systems, and the exploration of processes such as DNA folding, RNA folding, and nucleic acid-protein interactions.

- *Intrinsically Disordered Proteins*: CG models have been utilized to study intrinsically disordered proteins (IDPs) and their dynamics. CG representations can capture the essential features and conformation ensembles of IDPs, providing insights into their function, binding mechanisms, and interactions with other biomolecules.
- *Self-Assembly and Supramolecular Systems*: CG models are valuable for investigating the self-assembly and behavior of supramolecular systems, such as peptides, polymers, nanoparticles, and micelles. These models allow for the study of large-scale assembly processes, the exploration of phase transitions, and the understanding of complex self-assembly phenomena.
- *Drug Design and Binding*: CG models are used in drug design and the study of ligand-protein interactions. By coarse-graining ligands and proteins, these models can explore binding pathways, analyze binding affinities, and aid in the design of potential drug candidates.

These are just a few examples of the diverse applications of CG models in MD. CG approaches offer the advantage of bridging the gap between atomistic simulations and experimental observations, providing insights into complex biological systems and processes that are otherwise computationally challenging or experimentally inaccessible.

All in all, CG models in MD continue to play a vital role in investigating the dynamics, assembly, and function of biomolecular systems. Their ability to capture essential features, efficiently sample conformation space, and provide insights into collective motions make them valuable tools for studying complex biological processes.

8.3 Computational Tools for Molecular Dynamics

The community of researchers utilizing molecular dynamics (MD) simulations for studying proteins has experienced significant growth over the years, with a remarkable surge in interest during the COVID-19 pandemic. This expansion can be attributed to several factors, including advancements in computational power, improved software tools, and the need to address pressing biological questions.

- *Advancements in Computational Power*: The continuous progress in high-performance computing (HPC) resources and the availability of supercomputers and high-performance clusters have enabled researchers to perform more extensive and complex MD simulations. This increased computational power has facilitated the study of larger systems, longer timescales, and more sophisticated analysis, attracting a broader range of researchers to the field.

- *Enhanced Software Tools and Accessibility*: The development and widespread availability of user-friendly MD simulation software packages have lowered the barriers to entry for researchers interested in utilizing MD techniques. These software packages often come with intuitive graphical interfaces, detailed documentation, and online resources, making it easier for researchers from diverse backgrounds to adopt and employ MD simulations in their work.
- *Expanding Knowledge Base and Methodological Advances*: The accumulation of experimental data and advances in theoretical understanding of protein structure, dynamics, and function have contributed to the increased adoption of MD simulations. Researchers can now incorporate experimental data into MD simulations more effectively, making them a valuable complement to experimental approaches. Moreover, the development of advanced simulation methodologies, such as enhanced sampling techniques and coarse-grained models, has further expanded the applicability of MD simulations.
- *Pandemics and Biological Urgencies*: The COVID-19 pandemic, in particular, has led to a surge in interest and research employing MD simulations. The urgent need to understand the behavior of SARS-CoV-2 viral proteins, their interactions with host proteins, and the development of potential therapeutics prompted many researchers to turn to MD simulations. The pandemic highlighted the crucial role of computational methods in providing insights into viral dynamics, drug design, and vaccine development.

To support this more and more crowded community, computational tools play a crucial role in carrying out molecular dynamics (MD) simulations of proteins, providing researchers with the necessary software and resources to study protein dynamics, structure, and function. In this section, we will discuss various computational tools commonly used for MD simulations of proteins. These tools encompass both software packages and online resources that facilitate different aspects of MD simulations.

MD simulation packages are essential tools for studying protein dynamics; they provide diverse force fields, simulation protocols, and analysis capabilities, enabling researchers to simulate and analyze biomolecular systems efficiently. Here as follows a list of the most popular packages within the research community:

- *GROMACS*: GROMACS is a widely used MD simulation package that provides a comprehensive suite of tools for simulating biomolecular systems. It offers high-performance parallel simulations and features a wide range of force fields and analysis modules.
- *AMBER*: AMBER (Assisted Model Building with Energy Refinement) is another popular MD simulation package that offers a range of force fields, tools for system setup, simulation, and analysis. It is widely used for simulating biomolecular systems, including proteins.
- *NAMD*: NAMD (Nanoscale Molecular Dynamics) is a parallel MD simulation package known for its scalability and performance on high-performance com-

puting (HPC) systems. It supports a range of biomolecular systems, including proteins, and offers various analysis tools.
- *CHARMM*: CHARMM (Chemistry at HARvard Molecular Mechanics) is a versatile simulation package that provides tools for MD simulations, energy minimization, and analysis. It offers a wide range of force fields and is commonly used for simulating proteins and other biomolecules.
- *OpenMM*: OpenMM is an open-source MD library that provides a high-performance platform for MD simulations. It allows for easy integration with various programming languages and offers flexibility in building customized simulations and analysis protocols.

Here is a list of some web servers available for carrying out MD simulations of proteins, along with a brief description of each:

- *MDWeb*: MDWeb is a user-friendly web server that provides a comprehensive platform for setting up and running MD simulations. It offers various force fields, simulation options, and analysis tools, allowing users to simulate and analyze protein dynamics easily.
- *HADDOCK*: HADDOCK (High Ambiguity Driven biomolecular DOCKing) is a web server primarily focused on protein-protein docking, but it also provides an MD-based refinement option. Users can submit protein complexes and refine them using MD simulations combined with energy-based scoring functions.
- *MedusaDock* : MedusaDock is a web server that offers a protein-protein docking approach based on flexibility and molecular dynamics. It performs conformation sampling of proteins and protein-protein complexes using MD simulations to explore potential binding modes.
- *BioSimSpace WebApp*: BioSimSpace WebApp is an online platform that provides access to various MD simulation tools and workflows. It offers a range of force fields, simulation protocols, and analysis modules, enabling users to carry out complex MD simulations easily.
- *MDsrv*: MDsrv is a web-based molecular dynamics visualization and analysis tool. It allows users to upload MD trajectories and visualize protein structures, analyze protein dynamics, and generate interactive 3D representations of the simulations.

These are just a few examples of web servers available for MD simulations of proteins. Each server has its own unique features, capabilities, and user interfaces, allowing researchers to access and utilize MD simulation tools without the need for extensive computational resources or expertise in setting up complex simulations.

To support the MD simulations, some tools are required for the proper visualization and analysis, such as the following:

- *VMD*: VMD (Visual Molecular Dynamics) is a powerful software for visualizing, analyzing, and animating biomolecular systems. It offers a range of visualization options, analysis plugins, and tools for preparing simulation inputs.

- *PyMOL*: PyMOL is a popular molecular visualization tool that provides a user-friendly interface for visualizing and analyzing protein structures. It offers a range of features for rendering, analysis, and creating publication-quality figures, such as several plugins for molecular structure analysis.
- *Bio3D*: Bio3D is an R package specifically designed for analyzing protein structure, dynamics, and sequence-related data. It offers a comprehensive suite of functions for various tasks, including trajectory analysis, protein structure comparison, and network analysis.

These are just a few examples of the many computational tools available for MD simulations of proteins. The choice of specific tools depends on the research question, system complexity, available computational resources, and user preferences. Researchers often utilize a combination of software packages, visualization tools, and online resources to carry out comprehensive MD simulations and analyze the resulting data.

Additional Resources and Recommended Literature

General references for protein MD simulations can be found in [1, 17].

Protein Structure Prediction and Analysis 9

Reality is created by the mind; we can change our reality by changing our mind.

Plato

Abstract

Why it is important to know this material?

Protein structure prediction is one of the hot topics of the modern computational biology. The computational approach to the determination of protein structure is a new benchmark in structural biology, with a special regards to proteins recalcitrant to crystallization.

What is the key idea?

The key idea is to exploit the method of molecular dynamics to determine the three-dimensional protein structure according to the minimum of the conformation Gibbs free energy.

What is necessary to know already?

It is necessary to be familiar with the principles of protein structure and folding, along with the fundamentals of molecular dynamics of biomacromolecules.

9.1 The Computational Approach to the Protein Folding Problem

The protein folding problem is central problem in the theory of biophysical chemistry. Max Perutz was able for the first time in 1959 to reach a model for the molecular structure of a protein, the myoglobin, opening the route to the central technique for crystallography, the X-ray diffraction. Since then, more than 150,000 molecular protein structures have been resolved through this method and stored in

L. Di Paola, *Fundamentals of Molecular Bioengineering*,
https://doi.org/10.1007/978-3-031-42022-1_9

the Protein Data Bank, probably the most important data-sharing platform in the scientific community. In total, this platform stores more than 200,000 molecular structures including protein moieties, experimentally resolved through different techniques (including NMR, cryo-EM, and neutron scattering). This plethora of molecular data, stored in a general format, asks for a detailed explanation, trying to find the answer to the questions: how the chemical composition affects the structure? Is it possible to predict the structure only on the basis of the chemical composition?

These questions would be very simple to answer in the field of the organic chemistry, but the complex nature of biochemical systems challenges a trivial solution.

A central task of the protein folding problem is to understand the building roles of protein structure, in a process called folding, which occurs in nature in an apparently unfavorable environment, highly crowded of macromolecular species, with a very low water activity, to define the chemo-physical properties.

The computational approach to the protein structure prediction transforms the building roles into simulation elements, for instance, posing constraints to the motion of chain motion. However, as in nature, this approach requires huge computational resources, and only recently, under reasonable hypotheses and within certain constraints, it has given reliable results for protein structure prediction.

In any case, either the protein structure has been experimentally determined or not, the stability prediction from the molecular structure is essential to evaluate different properties of the protein molecular structures, how they can, for instance, be conserved for a long time in drug products or the effect of residues mutations on protein stability and in vivo turnover, for instance.

9.1.1 The Theory of Protein Folding

Since the first resolved protein molecular structure in 1959, a major scientific question emerged: which are the building roles of protein three-dimensional structure?

As previously seen, the protein molecular structure is organized in four hierarchical levels: the primary structure is what roughly can be addressed to as the "chemical composition" of the protein molecules (its sequence of amino acids, as they are combined from the amino to the carboxyl terminals). The term *folding* refers to the process of formation of the protein native conformation, which exerts the correct function in its physiological environment.

In 1973, Christian Anfinsen was awarded of the Nobel Prize for his outstanding contribution to understanding the building rules of protein structures. He established a thermodynamic hypothesis for the active protein structure (native structure) formation on the basis of studies on the renaturation (folding upon denaturation) of ribonuclease.

The fact the native structure was reached again after removing the chemical denaturant from the protein environment led him to state that the protein structure is solely determined by its chemical composition and the resulting interactions it determines with the environment, guiding the protein folding toward a unique, functioning three–dimensional structure.

The Anfinsen experiment leading to the thermodynamic hypothesis was based on the denaturation of ribonuclease by urea + mercaptoethanol produces an unfolded random coil. Reversing the denaturation conditions (removing the chemical denaturants) produces a folded structure where disulfide contacts oxidate again in the correct way.

The explanation leads to the thermodynamic hypothesis; citing the original words of Anfinsen in his Nobel lecture, "*This hypothesis states that the three-dimensional structure of a native protein in its normal physiological milieu (solvent, pH, ionic strength, presence of other components such as metal ions or prosthetic groups, temperature, etc.) is the one in which the Gibbs free energy of the whole system is lowest; that is, that the native conformation is determined by the totality of interatomic interactions and hence by the amino acid sequence, in a given environment. In terms of natural selection through the "design" of macromolecules during evolution, this idea emphasized the fact that a protein molecule only makes stable, structural sense when it exists under conditions similar to those for which it was selected - the so-called physiological state.*" The very base of the *Anfinsen's theory* is that the protein structure at both local and global level is only result of the physicochemical interactions between the protein molecular structure and the solvent environment. Stable protein structures correspond to favorable intramolecular interactions between residues, providing the conformation and the energy reservoir to promote the macromolecular structure solvation. The ... *Gibbs free energy of the whole system...* mentioned by Anfinsen is the Gibbs free energy conformation energy in the configuration space. This energy surface is usually referred to as *free-energy landscape*, addressing its ruggedness, corresponding to the multitude of possible conformations the protein molecular structure can assume (see Fig. 9.1). In general, the conformation energy, determined by the interactions between the residues in the molecule and between the protein molecular structure with the solvent, depends on temperature: for this reason, the energy is expressed in terms of thermal energy $k_B T$.

The ruggedness of the conformation energy corresponds to many local minima, which correspond to locally stable conformations (for small conformation variations around the minimum). Only one of these minima, not necessarily the absolute minimum, corresponds to the native structure, active at physiological conditions. The other conformations corresponding to local minima can be only partially active or inactive and can also cause negative effects (such as protein aggregates, whose physiological turnover is largely impaired or absent, so they tend to accumulate with associated tissue damage). In other words, these alternative conformations, nonnative, are also *kinetic traps* in the folding process, meaning that when the protein explores the conformation space, it can be "trapped" in one of the minima and so be hindered to reach the correct, native conformation.

The thermodynamic hypothesis addresses to the molecular interactions encoded in the residue composition the only cause of the native structure formation, denying any possible effect of the kinetic routes. However, some protein science topics

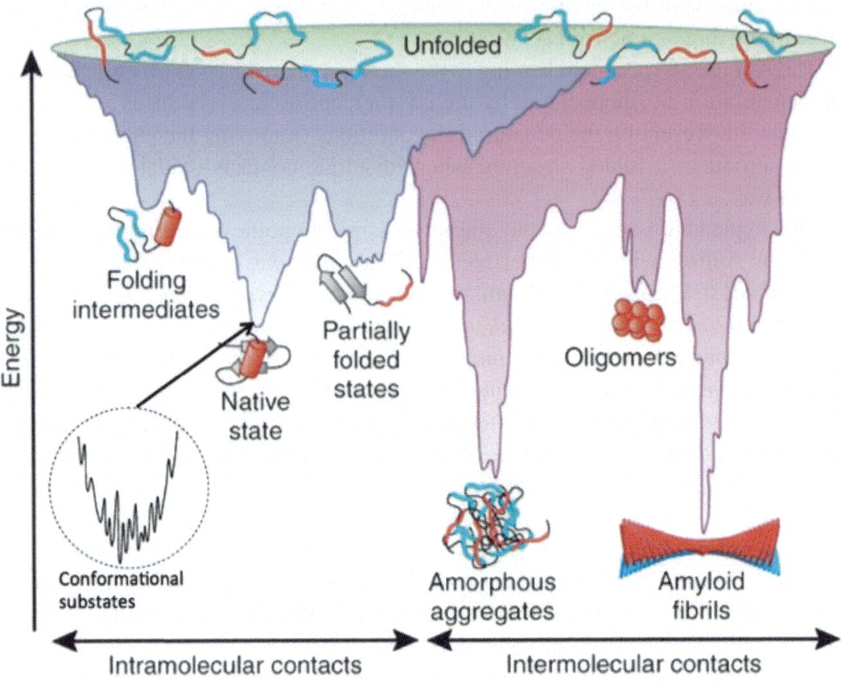

Fig. 9.1 The free energy conformation landscape of a protein: native structure occupies in general a local minimum in the rugged conformation energy surface, while other conformations—such as misfolded or aggregated forms—lie in similar energy traps. Reprint with permission from [35]

emerged more recently have partially confuted the Anfinsen's hypothesis, such a steady pillar in protein science for a long time to be casually known as *Anfinsen's dogma*.

However, recent observations and emerging topics in the protein science have partially confuted the Anfinsen's dogma, which well applies to the reversible unfolding of small single chain protein molecules. The recent observations deal mainly with: (a) the presence of few, yet not one, native conformations that are functionally active; (b) protein misfolding, leading to wrongly folded conformations whose aggregates are linked to neurodegenerative diseases (Parkinson's and Alzheimer's diseases).

In a protein of N amino acids, in a coarse grain approach—i.e., collapsing the whole residue into the position of the α carbons—the degrees of freedom are $3 \cdot N$. However, the geometrical constraints posed by the peptide bonds and by the sterical hindrance of side chains strongly reduce the possible conformation and, so, the degrees of freedom. These geometrical constraints corresponding to repulsive interactions, hindering not allowed possible conformations. The size of the configuration ensemble is so strongly reduced, but still huge. Additionally, the conformations corresponding to minima of the Gibbs free energy landscape are several, and only one of these is the native conformation.

According to all these issues, posing possible arguments against the Anfinsen's thermodynamic hypothesis, Cyrus Levinthal formulated in 1969 a mind experiment: a protein of 100 amino acids can assume around 10^{70} different conformations. If the conformation space is explored randomly by the protein polypeptide chain, we can expect very long times for protein folding, much larger than those observed and compatible with life. Thus the folding kinetics should be dominated by nonrandom processes, where the conformation evolution toward the minimum of the Gibbs free energy is guided through the energy landscape. The*funnel landscape theory* states the energy landscape is shaped as a funnel, and conformation dynamics follows pathways along this rough funnel surface Thus the exploration of the energy landscape is not unbiased, but the protein molecule explores the landscape in a biased way, following given paths leading swiftly to the local minima.

This concept is also summarized in the *minimum frustration principle*, outlined by Wolynes in terms, "*which quantifies the dominance of interactions stabilizing the specific native structure over other interactions that would favor nonnative, topologically distinct traps. In other words, the energy landscape of evolved proteins appears to be funneled*."

The kinetics of folding depends on the exploration of this conformation ensemble, and the transition from one conformation to another is faster the lower the jump required by the molecule to pass from one conformation hole to another.

Additionally, as previously stated, the transition is not random but follows the paths along the funnel, making much faster finding the native pit.

The funnel landscape of protein folding and the minimum frustration principle also outline the evolutionary path of protein molecular structures, being the native structure the final goal addressing the biological function, in most case through a folded conformation.

In the very last years, the research about protein folding kinetics has shed light on key aspects, such as the following:

- *The role of intermediates*: proteins often form intermediate structures during the folding process that play a critical role in determining the final protein structure. The analysis of these intermediates helps understanding the mechanisms of the protein folding and how the process can be influenced.
- *Characterization of the energy landscape*: recently, new methods have emerged to characterize the energy landscape of proteins, including the use of simulation techniques and experimental methods such as NMR spectroscopy. These methods also help to understand the thermodynamics and kinetics of the folding process and to identify key factors that influence the protein folding kinetics.
- *the role of molecular chaperones*: large proteins often require assistance from molecular chaperones, molecular machines of a protein nature that support and guide the correct folding. These proteins act as "folding catalysts" that help to accelerate the folding process and prevent misfolding. Understanding the role of the molecular chaperons into the folding process also reveals additional elements

to its mechanism and, more in general, the role of these factors in cellular physiology.

- *the role of co-translational folding*: proteins often begin to fold, while they are still being synthesized, and that the folding process is often influenced by factors such as the local environment of the ribosome, the presence of other proteins, and the rate of protein synthesis. The interaction of the nascent peptide with the ribosome in a macromolecular crowded environment results into the formation of intermediates and folding patterns that are not observed when the folding happens in solution. For this reason, it is crucial to study and understand the folding in the environment where it really occurs, to predict the in vivo folding mechanism.

All in all, the folding complex is a typical behavior of a complex system, adopting strategies to select given interactions between residues, i.e., those stabilizing the folded native structures. This complex nature of the protein molecular structures is a key property explaining as well their adaptive behavior: protein molecular structures work as adaptive elements able to feel the environment cues.

In this perspective, protein folding and binding are strongly connected and cannot be analyzed and solved separately, as discussed aptly in the protein-binding Chaps. 7 and 11. The complex systems theory, altogether, provides a general framework for both, shedding unexpected light on the structure-function protein problem, as described in the following chapter.

9.1.2 Werewolves, Shapeshifters, and Replicants: Strange Figures Among the Protein Community

The canon "*one sequence, one structure, one function*" has been disrupted by different exceptions leading to a more varied scenario including molecular structures endowed with a huge flexibility.

Yet, the studies of molecular structures through NMR have provided a general perspective of quite flexible protein molecular structures, which are able to use conformation changes to adapt to environment. Thus, each protein molecule exists in a wide conformation ensemble, whose single possible conformations correspond to local energy minima (molecules in the pits in Fig. 9.2).

In this complex scenario where proteins adopt multiple conformations near the native structure determined by the energy well, some peculiar groups of protein molecular structures are observed. These "freaks" are characterized by an anomalous conformation energy landscape, which also corresponds to abnormal functional properties.

The first class is represented by the *natively unfolded proteins*: these proteins do not have a stable folded structure in physiological conditions but reach a folded stable state when they bind their ligands (typically, of macromolecular nature).

Unlike folded proteins, which have a single or a few low-energy conformations, natively unfolded proteins exist in a highly populated, energetically disordered

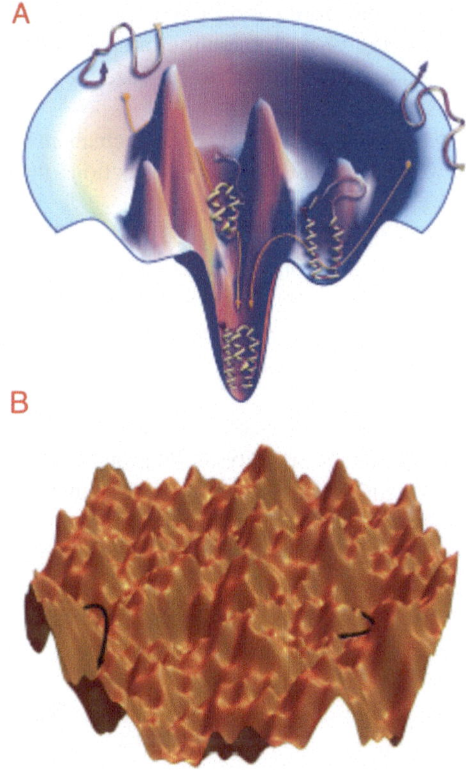

Fig. 9.2 The conformation energy landscape is very different for natively unfolded proteins with respect to native folded proteins. (**a**) Typical funnel energy landscape for folded native protein structures; (**b**) the energy landscape for natively unfolded proteins resembles that of glassy solids: no unique potential well appears and a highly rough surface presents a huge number of local minima. Reprinted with permission from [38]

ensemble of conformations. The energy landscape for natively unfolded proteins is no more clearly funneled, with many similar conformations of relatively low energy, corresponding to shallows potential wells.

This lack of a clear energy well means that natively unfolded proteins are not constrained to a single conformation and are able to adopt a wide range of structures. The absence of a stable, well-defined structure allows natively unfolded proteins to perform functions that require flexibility and adaptability, such as molecular recognition and regulation of cellular processes.

Figure 9.2 reports the comparison between the conformation energy landscape for a folded native conformation (panel A) and a natively unfolded protein structure. The strongly rugged surface does not have a unique potential well but a wide number of shallow potential wells.

Natively unfolded proteins can reach a stable structure upon binding by exploiting the binding energy to drive the formation of a stable, three-dimensional structure. This process is known as *induced folding*.

When a natively unfolded protein interacts with a ligand (whether macromolecular or not), the binding energy gain can drive the protein to reach a well-defined, stable conformation. This can occur through a number of different mechanisms, including the formation of specific hydrogen bonds, the creation of hydrophobic interactions, or the generation of specific salt bridges between the two molecules. The binding interaction can stabilize the protein and prevent it from adopting other, less stable conformations.

Induced folding is a key mechanism by which natively unfolded proteins can carry out their biological functions through specific interactions with ligand molecules.

The stability of the induced fold can be further stabilized by additional interactions with chaperones or other cellular components, allowing the protein to perform its function for a longer period of time.

More in general, binding can either increase or decrease the stability of a protein, and the impact on stability depends on the specific type of binding and the protein involved.

For example, in the case of transport proteins, binding to the transported substrate can increase the stability of the protein. This is because the binding often stabilizes the protein's structure and can prevent it from denaturing. The stability increase simply shifts over the complex formation, which has to stay stable for a long time (i.e., the hemoglobin—oxygen complex must last until blood reaches tissues requiring oxygen).

On the other hand, binding can also destabilize a protein, particularly in the case of enzymes where binding to a substrate can cause a conformation change in the protein that makes it more susceptible to denaturation. This is particularly the case of the formation of an unstable intermediate complex enzyme substrate, which fast transforms into the free state of the enzyme plus the reaction products.

Moonlighting proteins exert different functions, and their role is crucial in many pathological and physiological processes. They are present in all organisms and exert a wide range of functions, such as enzymes, transcription factors, chaperones, and so on. The combination of different functions is due to the presence of an active site combined to allosteric spots, where the interaction with other proteins is possible.

Another possible mechanism is that large conformation changes allow the formation of different active sites in different regions of the moonlighting protein. Finally, in this survey of freak proteins, the replicant misfolded proteins are gaining a day-by-day growing role in degenerative diseases, from neurodegenerative, Alzheimer's and Parkinson's diseases, to endocrinological, diabetes, and cancer, misfolding of p53.

The native conformation is often only metastable, occupying a local minimum in the energy landscape. Often this new conformation, referred to as *misfolded*, shows a lower solubility than the native one, producing the irreversible formation of aggregates that accumulates in tissues, provoking their loss of function and necrosis.

These aggregates are often referred to *amyloids*, the abnormal fibrous, extracellular deposits of protein aggregates found in organs and tissues. These deposits are the etiological agents of misfolding diseases.

From a thermodynamic point of view, these aggregates are very stable, often way more stable than the native conformations, so their formation can be considered irreversible.

A specific subclass of amyloids, *prions*, also possess the ability to bind other similar proteins in their native conformation and transform them in the misfolded conformation. For this reason, they are addressed to as "infectious" proteins associated with neurodegenerative diseases (transmissible spongiform encephalopathies—TSEs).

From this fast review of possible anomalous structure-function relationship in protein molecular structures, it is evident there is no simple and unique relationship. It is rather evident the stability of the protein structures is questionable insofar as its flexibility and quite large conformation fluctuations are required for function.

All in all, this wide variety of protein species is due to the roughness of the energy landscape, allowing many possible metastable conformations of the same protein structure, leading not only to pathological outcomes, as in the case of prions, but also providing for the adaptability of the protein molecular structures to the environment stimuli.

9.2 Computational Approaches and Tools for Protein Folding Prediction

In recent years, the development of high-performing computing platforms and the growing availability of cloud-based resources has revolutionized the field of protein folding prediction in recent years, making possible simulations and calculations once impossible. For example, Molecular Dynamics simulations, once limited to small proteins over short timescales, can now be used to simulate large protein complexes over microseconds or even milliseconds. This allows researchers to explore the conformation space of a protein more comprehensively and predict its native structure with greater accuracy.

On the other hand, the availability of large-scale datasets, such as the Protein Data Bank, has enabled the development of machine learning algorithm for predict protein folding: through the high-throughput analysis of thousands protein structure, it is possible to identify patterns and predict the structure of new proteins with high accuracy.

The growing relevance of computational approaches to protein folding has also led to the development of new tools and software packages that can be used by researchers without extensive computational expertise, thanks to user-friendly interfaces. This has made protein folding prediction more accessible to a broader range of researchers and has led to a more collaborative and interdisciplinary approach to protein research.

There are several different methods for the computational protein solving, which roughly can be grouped into two classes:

- *non-heuristic methods*: in this category fall different methods that are grounded on a physical representation of all forces and interactions contributing to protein folding, such as the following:
  - the **Molecular Dynamics** (MD) method, based on a physical representation of the interactions between residues in the protein folding problem;
  - **ab initio** methods, which predict protein folding from basic physical principles, without relying on any prior knowledge of protein structure;
- *heuristic* methods: these methods include all approaches based involve simplification and approximation of protein folding, such as the following:
  - **CG** modeling: this method simplifies the protein representation by focusing only on some atoms of the protein structure model, leading to faster simulations. However, this approximation can also lead to a loss of accuracy.
  - **fragment assembly** methods: this method assembles short segments of known structures to form a complete protein structure. This approach is based on the hierarchical and modular character of the protein folding processes, which proceeds by construction of small blocks (secondary structures) later assembled in more complex three-dimensional structures. This approach is failing when large conformation rearrangement occurs in large structures upon tertiary and quaternary structure formation;
  - **knowledge-based** methods: these methods use statistical potentials derived from known protein structures to predict the native conformation of a protein of unknown structure by comparing the sequence similarity with known structure. The basic assumption underlying these methods is that the larger is the sequence similarity the larger is the similarity in three-dimensional structure. These methods are also addressed to as *homology modeling* and will be thoroughly explained later.
  - **machine learning** methods: these methods are based on algorithms trained on a large dataset of known protein structures. While these algorithms can improve the accuracy of predictions, they also rely on the assumption that the dataset is representative of all possible protein folding patterns.

In recent years, many desktops have made available by different provider, among which:

1. **Rosetta**: widely used suite of protein modeling software that can predict protein structures, protein-protein interactions, and protein design. It is based on different computational methods, such as MD simulations, Monte Carlo simulations, and knowledge-based potentials, and has been tested and demonstrated to be highly accurate;
2. **I-TASSER**: a web-based tool based on iterative threading assembly refinement to predict protein folding, combining modeling, and MD simulations to accurately generate protein models;
3. **AlphaFold**: a deep learning-based method developed by DeepMind adopting the high capacities of cloud computing to predict protein structures with high accuracy, driving a significant improvement over previous methods;

4. **Modeller**: a popular protein modeling software combining homology modeling to generate protein models in a high accurate way.
5. **CASP**: CASP(Critical Assessment of Structure Prediction) is a massive community-wide, worldwide experiment for protein structure prediction, structured as a competition aimed at testing the best tools and showing the most significant results in protein structure prediction. The competition takes place every 2 years since 1994;
6. **AlphaFold2**: AlphaFold 2, in 2020, in the full pandemic period, won CASP14, marking an outstanding improvement of the accuracy of the computed three-dimensional protein structures.

During the pandemics, the proteome structure of SARS-CoV2, the virus responsible of the COVID-19 pandemics, was modeled by I-TASSER few weeks after the genome definition (see Fig. 9.3, on the basis of the structural information about the proteome of similar CoV (SARS and MERS). This is a bright example of the great potentiality of all prediction methods of protein folding, providing accurate estimation of proteins of unknown structure, which in turn is crucial to determine protein function and design purposed therapeutic strategies.

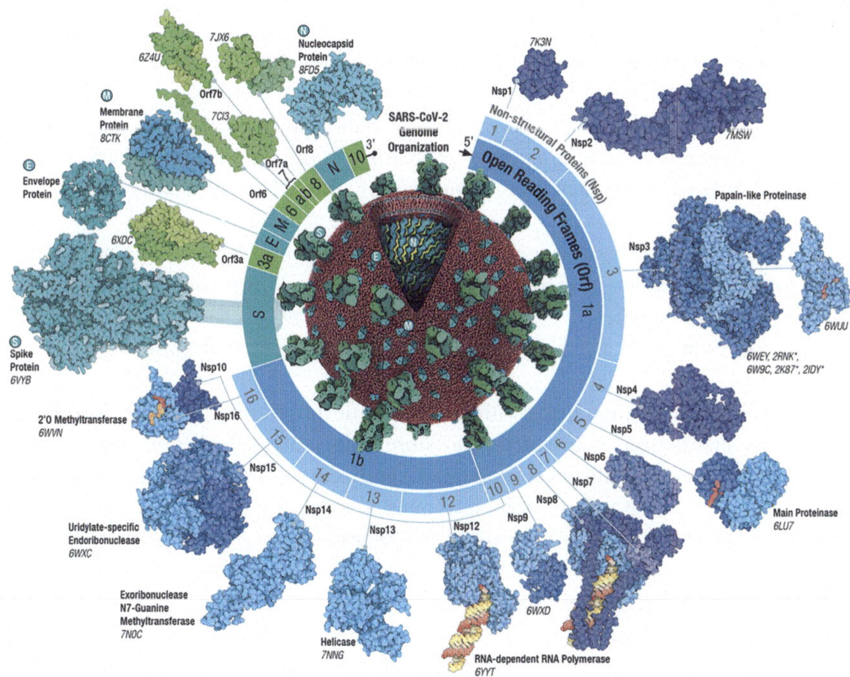

Fig. 9.3 SARS-CoV2 proteome. From the Protein Databank website

9.3 Sequence Alignment and Homology Modeling

Sequence alignment of proteins and nucleic acids is a key tool in biology, central in fields from evolutionary biology to structural biology. The sequence alignment is aimed at finding sequence similarities arising from shared ancestors, where single point mutations, insertions, and deletions help understanding the molecular path of evolution.

Sequence alignment of protein residues is important, because proteins with similar sequences are likely to have similar folds. This means that detecting large similarities in sequence between a protein with an unknown structure and a protein with a known structure can be an effective starting point for determining the unknown structure quickly, using homology modeling.

This incremental approach to the prediction of protein structure has strongly accelerated the determination of protein structure in a whole-proteome scale, for instance, as previously mentioned, as in the case of the very fast assessment of the structural biology of SARS-CoV2 proteome (see Fig. 9.3), months before the first experimental determination of the structure of key proteins was produced. This strongly accelerated the structure-based research about therapies and vaccines for the virus.

9.3.1 The Protein Sequence Alignment Problem and Its Applications

Protein sequence alignment is a central problem in computational chemistry; it deals with the comparison of two or more sequences of amino acids to identify similarities and differences.

The sequence alignment for proteins is similar to nucleotide sequences alignment, yet more complex, due to the composition of proteins made up of 20 residues with different frequencies.

The most commonly used algorithms for protein sequence alignment are:

- the **Needleman-Wunsch** algorithm: a dynamic programming algorithm, based on the construction of a matrix that stores the scores of all possible pairwise alignments between the two sequences. The score of each alignment is determined by a scoring matrix, which assigns a score to each pair of amino acids based on their likelihood of occurring together in nature (i.e., as in the BLOSUM matrix, based on the frequencies of amino acid substitutions observed in a database of aligned protein sequences);
- the **Smith-Waterman** algorithm: it is similar to the Needleman-Wunsch algorithm but finds the optimal local alignment between two sequences rather than the global alignment. It is useful and effective in the case of sequences sharing regions of high similarity surrounded by regions of low similarity or gaps.

- **multiple sequence alignment** methods allow to align more than two sequence simultaneously. They are typically based on *iterative refinement algorithms*, such as ClustalW or MUSCLE, which progressively improve the alignment by iteratively refining the scoring matrix and adjusting the alignment based on the new matrix.

The accuracy of protein sequence alignment depends on several factors, such as the quality of the scoring matrix, the choice of gap penalties, and the evolutionary distance between the sequences being aligned. In some cases, manual adjustment of the alignment may be necessary to correct errors or ambiguities in the algorithmic alignment.

One of the main issues in sequence alignment is the problem of *gaps*, which occur when one sequence has an insertion or deletion relative to the other sequence. Gaps can arise due to errors in sequencing or assembly, gene duplication or deletion, or other evolutionary processes. Gaps can be accounted for in sequence alignment algorithms by introducing a *gap penalty*, which is a penalty for inserting or deleting a character in one of the sequences.

The main challenge in sequence alignment is to determine the optimal alignment between two sequences, which is the alignment that maximizes the similarity between the sequences. The optimal alignment is typically found by dynamic programming algorithms, such as the Needleman-Wunsch algorithm or the Smith-Waterman algorithm, which can be applied to global and local sequence alignments, respectively.

Overall, sequence alignment is a critical problem in computational biology that underlies many different applications. While sequence alignment algorithms have made significant progress in recent years, there are still many challenges to be addressed, particularly in the context of increasingly large and complex datasets.

In this perspective, the use of new AI technologies, such as *deep learning*, represents an outstanding platform to strongly improve the accuracy and the speed of protein sequence alignment, especially as for the multiple sequence alignment comparing a large number of sequences simultaneously.

There are several web-based tools available for protein sequence alignment, among which:

- **Clustal Omega**: a fast and accurate tool for multiple sequence alignment, aligning sequences through a combination of progressive alignment and HMM-based methods; due to its efficiency, it is suitable to handle huge datasets;
- **MUSCLE**: a widely used tool for multiple sequence alignment, it adopts an iterative refinement approach to progressively improve the alignment; it can handle large datasets and can produce high-quality alignments for both closely related and distantly related sequences;
- **T-Coffee**: a tool combining multiple sequence alignment methods, including progressive alignment, consistency-based alignment, and template-based alignment. T-Coffee is particularly effective for aligning distantly related sequences;

- **MAFFT**: a tool based on a fast Fourier transform algorithm to efficiently align large datasets of protein sequences. MAFFT can handle both global and local alignment and has been shown to be highly accurate and robust on a wide range of sequence types;
- **HMMER**: a tool for protein sequence alignment that uses hidden Markov models (HMMs) to identify conserved protein domains and motifs. It is particularly useful for identifying functional domains in proteins and can handle large datasets.
- **PSI-BLAST**: a tool for iterative sequence alignment that uses position-specific scoring matrices (PSSMs) to improve the quality of the alignment over multiple iterations. It is particularly useful for identifying distantly related homologs and can be used to search large sequence databases.

The choice of tool depends on the specific needs of the user and the nature of the sequences being aligned. It is important to evaluate the accuracy and performance of different tools before selecting the most appropriate one for a given application.

9.3.2 Computational Tools for Homology Modeling of Protein Structure

The Anfinsen declaration "*one sequence, one structure*" has guided protein structure prediction methods by relying on the relationship between sequence and structure. Additionally, the large number of possible protein folds suggests that predicting structures can be simplified by recognizing them within certain categories. This approach is especially effective when predicting the structure of proteins with unknown structure by comparing their sequence similarity with those of proteins with known structure. Accordingly, the prediction of new structures can be significantly simplified by refining them with respect to proteins that are closely related in sequence and have known three-dimensional structures.

In this framework, *homology modeling*, also known as *comparative modeling*, is a computational method for predicting the three-dimensional structure of a protein based on its amino acid sequence and the structure of a related protein with a known structure. Homology modeling relies on the assumption that proteins with similar sequences have similar structures and can be used to predict the structure of a protein that has not yet been experimentally determined.

Based on several assumptions and simplification, the homology modeling falls among the heuristic methods. This is because it relies on several assumptions and approximations to predict the structure of a protein based on a related protein with a known structure.

Homology modeling assumes that proteins with similar sequences have similar structures and that the structural and functional properties of the template protein are similar to those of the target protein. Additionally, homology modeling involves approximations in the modeling process, such as the use of scoring matrices to evaluate the quality of the sequence alignment and the use of energy minimization methods to refine the model.

While homology modeling is a powerful tool for protein structure prediction and has been shown to be accurate in many cases, it is still subject to limitations and uncertainties, and the accuracy of the resulting models is dependent on the quality of the template protein and the sequence alignment.

Protein molecular threading, also known as fold recognition, is a computational method for predicting the three-dimensional structure of a protein based on its amino acid sequence and the structures of related proteins with similar folds. This approach is similar to homology modeling in that it relies on the assumption that proteins with similar folds have similar structures and can be used to predict the structure of a protein that has not yet been experimentally determined.

However, there is an important difference between homology modeling and molecular threading. Homology modeling relies on the assumption that proteins with similar sequences have similar structures, whereas molecular threading focuses on the similarity of the overall fold, based on the abovementioned large degeneracy of few possible fold patterns in the light of the huge possible number of sequences. This means that molecular threading can be used to predict the structure of proteins that are not necessarily homologous to any known protein but have a similar fold to a protein with a known structure.

The basic principle of molecular threading is to identify proteins with similar folds in a protein structure database and then align the amino acid sequence of the target protein to the structural template using a scoring function that evaluates the compatibility between the target sequence and the template structure. The aim is to find the best-fitting alignment, corresponding to highest value of the scoring function, which can then be used to generate a model of the target protein.

Molecular threading has several advantages over other methods for predicting protein structure, such as ab initio modeling, which does not rely on a template structure. Molecular threading is generally faster and more accurate than ab initio modeling, especially when a suitable template protein is available.

However, molecular threading also has limitations. Like homology modeling, molecular threading accuracy depends on the quality of the alignment and the template structure, and errors in either of these can lead to inaccurate models. Additionally, molecular threading can only be used to predict the structure of proteins with similar folds to known proteins and may not be applicable to proteins with highly novel folds. Nonetheless, molecular threading is a valuable tool for predicting protein structures and has been used to successfully predict the structures of many proteins with unknown structures.

Homology modeling methods all involve the following steps:

- **Template selection**: To begin homology modeling, the initial task is to find a template protein with a known structure that is homologous to the target protein. This involves selecting a template that has a high sequence identity with the target protein, which indicates that they have a similar fold and function. The choice of template is typically based on shared functions and the fact that they belong to the same category of protein structures;

- **Sequence alignment**: The amino acid sequence of the target protein is aligned with the amino acid sequence of the template protein to identify regions of the target protein corresponding to regions of the template protein with a known structure.
- **Model building**: Using the sequence alignment as a guide, a model of the target protein is constructed by using the known structure of the template protein as a framework, with the matching regions of the target sequence corresponding to the structure of the template;
- **Model refinement**: After building the initial model, the final step in homology modeling is to enhance its accuracy and reliability. This can be achieved through model refinement techniques such as energy minimization, loop modeling, and side chain optimization, which help to improve the alignment of the model with the experimental data.

Homology modeling has several advantages over other methods for predicting protein structure, such as ab initio modeling, which does not rely on a template structure. Homology modeling is generally faster and more accurate than ab initio modeling, especially when a suitable template protein is available. Homology modeling can also provide insights into the function and evolution of proteins as well as help design experiments to test the predicted structure and function.

However, homology modeling also has limitations. Homology modeling is only as accurate as the quality of the sequence alignment and the template structure, and errors in either of these can lead to inaccurate models. Homology modeling is also less accurate for proteins with low sequence identity to known structures or with unusual folds. Finally, homology modeling cannot predict structural features that are not present in the template structure, such as posttranslational modifications or alternative conformations.

Although similar, the two methods show distinct differences, listed below:

1. **Template selection**: In homology modeling, a template protein with a known structure that is homologous to the target protein is selected based on sequence similarity and structure similarity. In threading, a set of template structures is selected from a protein structure database based on their compatibility with the target protein sequence.
2. **Model building**: In homology modeling, a model of the target protein is built by using the known structure of the template protein as a scaffold, with the matching regions of the target sequence corresponding to the structure of the template. In threading, the target protein sequence is threaded through each template structure to identify the best fit, and a model is built based on the optimal threading solution.
3. **Level of accuracy**: Homology modeling is generally considered more accurate than threading, because it relies on a higher degree of sequence and structure similarity between the target protein and the template protein. Threading, on

the other hand, is less accurate, because it uses a more approximate method for matching the target sequence to the template structure.

4. **Applicability**: Homology modeling is typically used for predicting the structures of proteins that are closely related to proteins with known structures, whereas threading can be used to predict the structures of proteins with no known structural homologs. Threading is particularly useful for predicting the structures of membrane proteins and other proteins that are difficult to crystallize.
5. **Computational complexity**: Homology modeling can be computationally expensive, because it requires a template protein with a known structure that is homologous to the target protein. Threading, on the other hand, is generally faster and more scalable, because it involves matching the target sequence to a set of template structures in a protein structure database.

Overall, homology modeling and threading are two different approaches for predicting protein structures, with different strengths and limitations. Homology modeling is more accurate but requires a closer homolog, while threading is less accurate but can predict the structures of proteins with no known homologs.

There are several web-based tools for the homology modeling, among which:

- **SWISS-MODEL**: this widely used tool for homology modeling allows users to generate protein models based on templates from the Protein Data Bank (PDB) or user-supplied templates. SWISS-MODEL uses a range of methods for model building and refinement;
- **Phyre2**: this tool adopts iterative threading assembly refinement to predict protein structures, generating models for sequences with no known structural homologs;
- **MODELLER**: this widely used tool generates homology models using multiple templates and a range of refinement techniques, including MD and energy minimization.
- **RaptorX**: this tool is based on a deep learning-based method for homology modeling called DeepTemplate. It is able to predict the structures of proteins with no detectable sequence similarity to known structures;
- **I-TASSER**: this tool generates protein models using combining threading, modeling, and MD simulations. It can generate models for sequences with no known structural homologs with high accuracy.

Both homology modeling and threading are expected to result in a substantial increase in the number of Computed Structure Models (CSMs), which, in turn, is likely to greatly enhance structure-based research for proteins that were previously not amenable to analysis because their crystallization is not feasible.

9.4 Computational Prediction of Protein Stability

The term *protein stability* refers to its ability to keep its folded structure under different environmental conditions and its proper functions altogether. On the other hand, deviation from the native state, due to environment factors—chemical and physical—and to mutations, can be the cause of several diseases and disorders. For these reasons, protein stability lies at the center of the stage in protein science theory and applications.

In drug discovery, protein stability is a key consideration in the design of protein-based therapeutics, as it can affect the efficacy, safety, and shelf life of the drug. In biotechnology, protein stability is important for the development of industrial enzymes and biocatalysts, as well as for the production and purification of proteins.

Computational prediction of protein stability has emerged as a powerful tool for protein science research and applications. By combining experimental measurements with computational modeling, it is possible to gain a more comprehensive understanding of protein stability and its underlying mechanisms. Accurate prediction of protein stability can facilitate the design and engineering of proteins with improved properties, as well as the discovery and development of new drugs and biocatalysts.

Overall, protein stability is a critical property of proteins that impacts many areas of protein science research and applications. Advances in computational modeling and experimental techniques have greatly enhanced our understanding of protein stability and will continue to drive progress in this important field.

9.4.1 Theoretical Description of Protein Stability

The thermodynamic definition of protein stability refers to the difference in energy between the native and denatured states of a protein. In the native state, a protein has a specific three-dimensional structure that is stabilized by non-covalent interactions, such as hydrogen bonds, hydrophobic interactions, and electrostatic forces.

The denatured state, on the other hand, is a disordered, unfolded state in which the protein loses its specific structure and interactions. The interactions stabilizing the folded state arise from the interaction of the residues with solvent, so hydrophobic residues repel the water contact, while hydrophilic residues are solvated due to an attractive interaction with the water molecules.

Thermodynamic stability is an important property of proteins that affects their folding, stability, and function. Proteins with low stability are more prone to denaturation and aggregation, which can lead to loss of function and disease. Conversely, proteins with high stability are less likely to undergo denaturation and are more robust under different environmental conditions.

The thermodynamic definition of protein stability has important implications for protein science research and applications. Understanding the factors that contribute to protein stability can facilitate the design and engineering of proteins with improved properties, such as increased stability, activity, or solubility. Accurate measurement and prediction of protein stability can also aid in drug discovery, biotechnology, and other areas of protein science research.

The stability of a protein is determined by the Gibbs free energy difference between the native and denatured states $\Delta G_{unf} = G_{unf} - G_{fold}$: a protein is considered stable if ΔG_{unf} is positive, indicating that the native state is thermodynamically favored over the denatured state (that means, the unfolding process is not thermodynamically favored). The module of ΔG_{unf} reflects the strength of the non-covalent interactions that stabilize the native state, with larger positive values indicating stronger interactions and greater stability.

As previously mentioned, protein solvation is another important factor affecting protein stability. Solvation refers to the interaction between the protein surface and water molecules in the surrounding environment. Water molecules can interact with protein surface residues through hydrogen bonding, electrostatic interactions, and van der Waals forces and can stabilize or destabilize the protein structure depending on the nature and strength of these interactions, which in turn depend on the chemical composition (primary sequence) of the protein.

Protein solvation is strictly linked to folding, as depicted in Fig. 9.4, where U_{NS} indicates the not solvated unfolded form, F_{NS} the folded not solvated (crystal), F_S the folded solvated form (native protein structure in water), and U_S the solvated

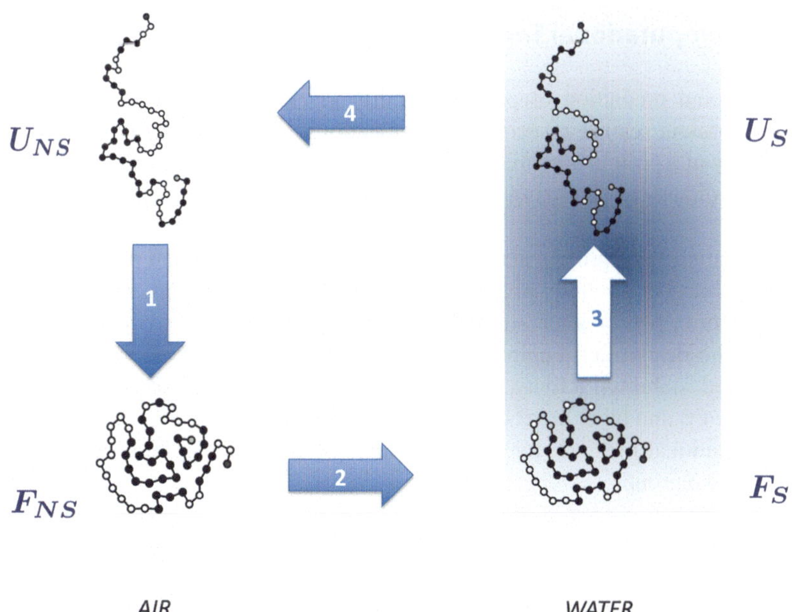

Fig. 9.4 Protein solvation affects folding through the protein solvation energy

unfolded. The solvation process is strictly step 2 in Fig. 9.4, while (un)folding in the cycle is the step 3.

Protein solvation can affect protein stability in several ways. Firstly, solvation can affect the conformation entropy of the protein, which refers to the degree of disorder or flexibility of the protein structure. Water molecules can interact with and immobilize specific protein residues, reducing the conformation entropy of the protein and stabilizing its structure. Conversely, water molecules can also disrupt protein-protein interactions and induce disorder in the protein structure, leading to decreased stability.

Secondly, solvation can affect the hydrophobic core of the protein, which is composed of nonpolar residues that are typically buried in the interior of the protein structure. Water molecules can interact with and disrupt the hydrophobic core, leading to destabilization of the protein structure. Conversely, water molecules can also interact with hydrophobic residues on the protein surface and stabilize the structure through hydrophobic interactions.

Finally, solvation can affect the kinetics of protein folding and unfolding. Water molecules can act as a barrier to protein folding by stabilizing partially folded intermediates or can assist in the folding process by facilitating the search for the native state, through a conformation energy optimization process. Changes in the solvation environment can therefore affect the rate and mechanism of protein folding and unfolding.

9.4.2 Computational Tools for Protein Stability Prediction

Computational prediction of protein stability given its structure is an important aspect of protein science research and applications. It is based on the principle that the stability of a protein is primarily determined by its structure and that the stability can be predicted by analyzing the structural features and interactions within the protein.

The prediction of protein stability is important, because it can provide insights into the stability and activity of a protein under different conditions, such as changes in temperature, pH, or ionic strength. Accurate prediction of protein stability can facilitate the design and engineering of proteins with improved properties, such as increased stability, activity, or specificity. It can also aid in drug discovery, biotechnology, and other areas of protein science research.

Computational methods for predicting protein stability can be broadly categorized into two groups:

- **physics-based** methods, such as MD simulations, are based on the definition of physical laws governing the behavior of a protein under different conditions and to calculate the thermodynamic properties of the protein based on its atomic-level interactions;

- **empirical** methods are based on the statistical analysis of experimental data and including all variables—sequence, structure, thermodynamic parameters—to predict protein stability.

One of the most widely used physics-based methods for predicting protein stability is based on molecular dynamics (MD) simulations. MD simulations involve the calculation of the forces and motions of atoms in a protein structure under different conditions and can be used to predict the stability of a protein under various environmental conditions, such as changes in temperature, pH, or ionic strength. MD simulations can also be used to identify the key structural features and interactions that contribute to protein stability and to guide the design of new proteins with improved stability and activity.

Empirical methods for predicting protein stability include statistical potentials, machine learning algorithms, and knowledge-based methods. These methods are based on the analysis of large datasets of experimental data on protein stability and use statistical models to predict the stability of new protein structures based on their sequence, structure, and thermodynamic features.

As such, computational methods for predicting protein stability based on solvation energy include molecular dynamics simulations and empirical methods, such as statistical potentials and machine learning algorithms. These methods use the solvation energy as a starting point to predict the stability of a protein under different conditions, such as changes in temperature, pH, or ionic strength.

MD simulations can be used to calculate the solvation energy of a protein structure under different conditions and to predict the effects of changes in the surrounding environment on protein stability. Empirical methods, on the other hand, use statistical models to predict the stability of a protein based on its solvation energy and other structural and thermodynamic features.

Starting from protein solvation energy in a physiological environment is a promising approach for predicting protein stability computationally, as it provides a direct measure of the interactions between the protein and the surrounding environment. By combining solvation energy with other structural and thermodynamic features, it is possible to develop accurate and robust models for predicting protein stability and guiding protein design and engineering.

There are many web tools to predict protein stability on the basis of its structure and sequence, among which:

- **DynaMine**, a web server that predicts protein stability combining MD simulations and machine learning algorithms to predict the changes in protein stability caused by changes in the surrounding environment, such as temperature and pH;
- **PROSS 2** a an automated structure and sequence-based design method for computing protein stability. The tool uses a statistical thermodynamic model to predict the changes in free energy of protein stability caused by changes in the surrounding environment.

Different web tools are available to predict protein stability on the basis of its sequence and structure to explore the variation of protein stability in terms of point mutations. In this case, the stability variation is given in terms, for instance, of $\Delta\Delta G_{unfold}$, that is, the variation of protein stability (in terms of ΔG_{unfold}) upon point mutations.

As follows, the most popular web server tools to estimate protein stability changes upon mutations:

- **FoldX**, a web server to predict protein stability changes upon mutations; the web server allows to submit a protein structure and a list of mutations, and it will predict the impact of the mutations on the protein stability.
- **PoPMuSiC** can predict the effects of point mutations on protein stability based on the thermodynamic cycle approach. Users can submit a protein structure and a list of mutations, and the server will provide the predicted changes in Gibbs free energy.
- **DUET** predicts the effects of point mutations on protein stability and protein-protein interactions. It uses an ensemble of machine learning models trained on experimental data to predict the effects of mutations.
- **SDM** predict the point mutations effect on protein stability based on the empirical energy function. Users can submit a protein structure and a list of mutations, and the server will provide the predicted changes in Gibbs free energy.

Web servers that predict the effects of point mutations on protein stability provide a valuable tool for protein engineering and drug design. These servers allow researchers to make informed decisions about which mutations to make in order to improve protein stability or function.

For example, a researcher may use a web server to predict the impact of a particular mutation on the stability of a protein that is being developed as a drug target. If the mutation is predicted to destabilize the protein, the researcher may choose to avoid that mutation and focus on other mutations that are predicted to improve stability.

Web servers that predict the effects of point mutations on protein stability can also be used to design new proteins with specific properties. For example, a researcher may use a web server to predict the effects of mutations on the stability of a protein that is being designed to have a particular enzymatic activity or binding affinity for a specific molecule.

9.4.3 Protein Stability Prediction in Quaternary Structure

The computational prediction of protein stability in multimeric complexes is crucial to determine the complexes functionality, which strongly depends on the complexes' stability, possibly be affected by mutations or changes in environmental conditions.

Several factors may influence the stability of protein complexes, including the interactions between the subunits, the packing of the protein interface, and the distribution of charges and hydrophobic residues across the interface. Predicting the stability of protein complexes requires an understanding of these factors and the ability to model the complex structure accurately.

Several computational methods have been developed to predict the stability of protein complexes. These methods include both empirical and physics-based approaches. Empirical methods rely on statistical analysis of known protein structures to derive rules that can be used to predict the stability of new complexes. Physics-based methods, on the other hand, use molecular mechanics or quantum mechanics simulations to calculate the Gibbs free energy of the complex and predict its stability.

One common approach to predicting the stability of protein complexes is to use protein-protein docking software to generate a model of the complex structure and then use energy-based methods to calculate the Gibbs free energy of the complex. This approach has been successful in predicting the stability of a wide range of protein complexes, including enzyme-substrate complexes, protein-DNA complexes, and antibody-antigen complexes.

Another approach to predicting the stability of protein complexes is to use machine learning methods to develop predictive models based on structural and sequence features of the protein subunits. These models can be trained on known multimeric protein complexes to predict the stability of new complexes.

There are diverse web server devoted to the analysis of multimeric protein complexes and compute their stability, among which:

- **RosettaDock**: a widely used web suite for protein structure prediction, design, and modeling. The Rosetta web server includes a module for predicting the stability of protein-protein interfaces in multimeric complexes.
- **PISA**: the Protein Interfaces, Surfaces, and Assemblies (PISA) web server can be used to analyze protein quaternary structures and predict their stability on the basis of the interaction energies between the subunits.
- **PDBePISA**: The PDBePISA web server is a related tool that is integrated with the Protein Data Bank in Europe (PDBe). PDBePISA can be used to analyze protein quaternary structures and predict their stability using a similar approach as PISA.
- **HADDOCK**: the HADDOCK web server is a protein-protein docking program that can be used to predict the structure and stability of multimeric complexes, taking into account the shape complementarity and electrostatic interactions between the subunits to predict the stability of the complex.
- **SwarmDock**: the SwarmDock web server is a protein-protein docking program used to predict the structure and stability of multimeric complexes, thanks to a swarm intelligence algorithm to search for the best-fitting complex structure and predict its stability.

Computing the stability of multimeric protein complexes is important for several reasons: protein multimeric complexes often play critical roles in biological processes, and their stability is crucial for their function. For example, protein-

protein interactions are involved in signaling pathways, enzymatic reactions, and DNA replication, and the stability of these interactions can affect the efficiency and specificity of these processes.

Second, protein complexes are potential targets for drug design. Many drugs target protein-protein interactions in order to disrupt protein complexes and interfere with biological processes (see, i.e., the drugs and vaccines targeting interactions in the complex of the SARS-CoV2 spike protein with its human receptor, the ACE2). By computing the stability of a protein complex, researchers can identify potential drug targets and design drugs that specifically disrupt these complexes.

Third, predicting the stability of protein complexes can guide protein engineering efforts. By understanding the factors that contribute to the stability of a complex, researchers can design mutations or modifications that improve the stability and function of the complex.

Finally, computational methods for predicting the stability of protein complexes can complement experimental methods. Experimental methods for determining the stability of protein complexes, such as isothermal titration calorimetry (ITC) or surface plasmon resonance (SPR), can be time-consuming and resource-intensive. Computational methods can provide a rapid and cost-effective alternative for screening potential protein-protein interactions and predicting the stability of protein complexes.

All in all, computing the stability of multimeric protein complexes is important for understanding their function, identifying drug targets, guiding protein engineering efforts, and complementing experimental methods. By predicting the stability of protein complexes, researchers can make informed decisions about drug design and protein engineering, ultimately leading to the development of new therapeutics and biotechnological applications.

Additional Resources and Recommended Literature

A very complete and comprehensive textbook is [19]. In addition, two reviews [31, 34] provide updated perspectives on the protein folding prediction.

Systems Biology and Structure-Based Protein Networks

10

In nature we never see anything isolated, but everything in connection with something else which is before it, beside it, under it and over it.

J.W. Goethe

Abstract

Why it is important to know this material?

The systems theory application to biology allows the cutting-edge understanding of biological mechanisms. The network approach to biomolecular systems analysis provides unique knowledge about the molecular machinery behind life.

What is the key idea?

The key idea is to recognize that interconnections between elements in a biological system play a key role in the system behavior.

What is necessary to know already?

It is necessary to have a general knowledge about systems theory, with a special regards to complex network theory and graph algebra.

10.1 Introduction

The system approach to biology roots in the first steps of this science, for instance, in the emerging evidence of the system nature, for instance, of human body provided by physiology study. The formalism of systems theory was the very basis to find its application in the biological sciences, at first with the interpretation of interactions of different species in a given shared physical space, indeed addressed to as *ecosystem*.

Starting from this concept, the complex systems theory has provided a comprehension of biological elements and their connections based on the observation of

L. Di Paola, *Fundamentals of Molecular Bioengineering*,
https://doi.org/10.1007/978-3-031-42022-1_10

biological elements sets as a whole. The general systems theory was formalized in the 40s of the twentieth century by Ludwig von Bertalanffy on the basis of the observation of the biological systems. The novelty of his approach was the identification of general concepts and properties of systems, where the definition of the single elements and the relationship between them are crucial to determine and interpret the properties of the system.

The systems biology today exploits the computational sources to develop sophisticated models to interpret a wide spectrum of biological phenomena, providing also priceless tools for predicting the dynamic behavior of biological entities. The systems biology contributes to the fine comprehension of complex behavior of biological systems, through predictive quantitative description of emergent properties and endeavors.

10.2 Basic Definition of Theory of Complex Systems

The notion of complex systems relates to sizable sets containing a substantial number of elements. When observed as a whole, these sets display sophisticated behavior that is not directly predictable by examining the individual properties of the elements alone. If it is entirely unfeasible to deduce the behavior of a set of elements by only considering the properties of the individual elements, this is a strong sign that the set constitutes a complex system.

The complex system behavior is primarily determined by the relationship between elements rather than by the properties of the elements. This leads to a strong generality of the approach, since very different systems, made up of very different elements, can share similar behaviors, if the structure of relationship between the elements is similar.

Complex systems exhibit several distinctive properties that set them apart from other systems. Some of the most relevant characteristics include the following:

1. **Emergence**: Complex systems often display emergent behavior, which means that the whole system exhibits properties or behaviors that cannot be predicted by simply examining the individual components.
2. **Adaptability**: These systems can adapt to changing environments or conditions by reorganizing themselves, evolving, or learning from experience.
3. **Nonlinearity**: Interactions within complex systems are often nonlinear, meaning that small changes in inputs or conditions can lead to disproportionately large effects on the system's behavior.
4. **Self-organization**: Complex systems tend to organize themselves without any central control, leading to the formation of patterns, structures, or behaviors.
5. **Interdependence**: The elements within a complex system are often interconnected and interdependent, meaning that the behavior of one element can significantly impact the behavior of others.
6. **Robustness**: Complex systems can be robust and resilient, maintaining their function and structure even in the face of disturbances or changes.

7. **Hierarchy**: These systems often exhibit hierarchical structures, where smaller subsystems are nested within larger ones, and each level has its own set of properties and behaviors.
8. **History and path dependence**: The behavior and properties of complex systems are often influenced by their history and past events, making them sensitive to initial conditions and subject to path dependence.
9. **Modularity**: Complex systems often exhibit a hierarchical organization, with modules or subsystems that can function independently or in coordination with other modules. This modularity allows for greater flexibility and adaptability;
10. **Interconnectivity**: Elements within complex systems are often highly connected, with many interactions occurring between components. This interconnectivity can lead to feedback loops and cascading effects that contribute to the system's overall behavior.
11. **Feedback loops**: They are situations where the output of a system feeds back into the system as input, leading to a chain of events that can either reinforce or counteract the original input. There are two types of feedback loops: positive and negative. Positive feedback loops can lead to explosive growth or collapse, while negative feedback loops can help maintain stability and balance. They are found in many complex systems, and understanding them is important for gaining insights into the underlying mechanisms that govern the behavior of these systems and developing strategies to promote stability and resilience.

All these properties are clearly ascribable to biological systems and, for this reason, modeling biological systems as complex systems provides deep insight on the behavior of biological entities.

10.2.1 Complex Systems and Networks

Networks are elective representations of complex systems, because they catch the essential feature of the central role of the relationships and interactions between the components of a system.

In the field of mathematics and computer science, a graph is a commonly used mathematical structure for representing a network. A *graph* is defined as a set of vertices (also known as nodes) connected by edges or links. The edges represent the relationships or connections between the vertices. Graph theory provides a framework for analyzing and modeling the behavior of networks.

While a graph is a specific mathematical structure, the term *network* is more general and can refer to any collection of interconnected components. In practice, networks are often represented using graphs, but other mathematical structures or models may be used to represent networks depending on the specific characteristics of the system being studied.

Networks can be used to represent a wide variety of complex systems, including social, transportation, and biological networks. For example, in a social network, the nodes might represent individuals, and the edges might represent friendships or professional relationships between them. Likewise, the nodes of a transportation network may represent cities or transportation hubs, and the edges might represent the roads, railways, or airline routes connecting them.

The use of networks as a mathematical representation of complex systems has several advantages. First, networks can be used to identify key components and interactions within a system. By analyzing the structure of a network, researchers can identify nodes with high centrality, which are critical for the functioning of the system.

Second, networks can be used to analyze the behavior of a system over time. By modeling the dynamics of a network, researchers can gain insights into how the system changes and evolves over time.

Third, networks can be used to predict the behavior of a system. By using network analysis techniques, researchers can predict the behavior of a system based on the relationships between its components.

When adopting a graph representation of the network, the mathematical descriptors of the graph directly describe the structure and function of the corresponding network.

The mathematical description of a network as graph requires the elements (nodes) are ordered. According to this, graphs are classified on the basis on the type of link connecting two nodes:

- **Undirected graph**: In this case, the edges do not have a specific direction and are represented by lines connecting the nodes. In general, undirected graphs represent networks where the connections between elements are reciprocal, like in the case of social networks, where the nodes represent people and the edges represent friendships. In this case, the relationship between two people is mutual, so there is no need to represent the direction of the relationship. Another application of interest is the molecular networks, where the nodes represent atoms and the edges represent chemical bonds between the atoms. In this case, the bonds are typically mutual, so there is no need to represent the direction of the bond. More in general, undirected graphs are useful for representing relationships where the direction of the relationship does not matter at all.
- **Directed graphs**: Each edge is represented by an arrow pointing from the source node to the target node. Directed graphs are useful for representing relationships where the direction of the relationship matters. For example, directed graphs can be used to represent networks of communication, such as email or phone calls, where each edge represents a communication from one person to another. The direction of the edge indicates who initiated the communication and who received it.
- **Unweighted graphs**: Each edge between two nodes has the same weight or cost, regardless of the length or strength of the relationship represented by the edge. These graphs are used to represent social networks, where the edges represent social relationships between people nodes: In this case, each relationship between two people is considered equal, regardless of its strength. Analogously, in molecular networks, the edges represent chemical bonds between atoms or

significant noncovalent intramolecular interactions between groups of atoms in macromolecules. In this case, the unweighted graph accounts only for a bond or interaction existing or not between two molecular nodes, regardless of the strength of bond or interactions. In general, unweighted graphs are useful for representing relationships where the strength or importance of the relationship is not important or unknown. They are often used as a simplification of more complex weighted graphs, where the strength or importance of each relationship is represented by a numerical value or weight.

- **Weighted graphs**: Each edge connecting two nodes is given of a *weight* that quantifies the intensity of the connection; in weighted graphs, the length or the strength of the connection matters. More in general, the weights can represent different things depending on the context of the graph, such as distance, cost, time, strength, probability, or similarity. Weighted graphs can be used to model a variety of real-world situations, where the edges represent the relationships or connections between entities, and the weights represent the strengths or distances between them. In molecular graphs, the weight can describe the strength of the interactions or bonds or the length of the same.

Figure 10.1 graphically represents the undirected and directed graphs, by comparison.

The quantitative description of protein topology passes by the definition of network descriptors , also called *invariants* , since independent from certain operations or transformations, such as the node relabeling.

Some of these descriptors are also defined as *centrality* measures, pointing to the fact that they describe the role of single nodes in the network topology.

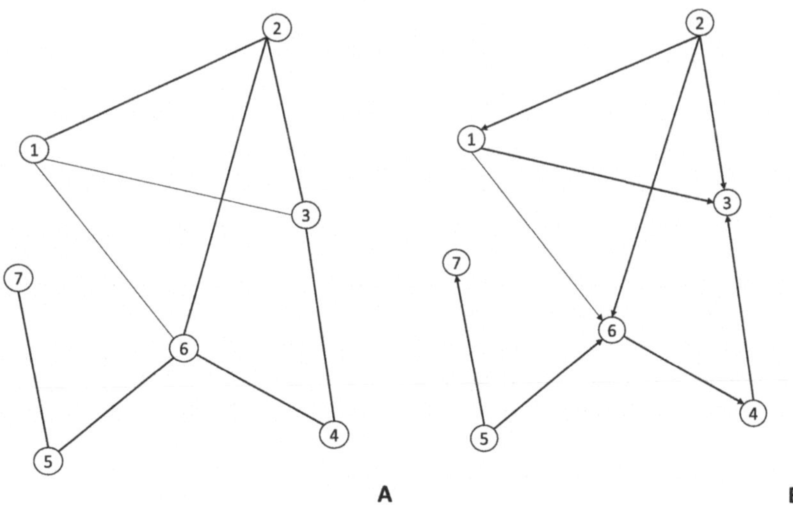

Fig. 10.1 Graphical representation of undirected (panel **a**) and directed graphs (panel **b**)

The simplest and the most important centrality metrics is the **node degree** or **degree centrality**. This measure is based on the number of connections a node has in the network, and it assumes that nodes with more connections are more central. In directed networks, there are two versions of this measure: in-degree centrality (based on the number of incoming connections) and out-degree centrality (based on the number of outgoing connections).

The **shortest path** between two nodes refers to the minimum number of connections (or edges) needed to travel from one node to the other in the network. This concept is closely related to the idea of distance between nodes, where the distance is defined as the length of the shortest path between the nodes.

The shortest path matrix is a matrix that contains the shortest path lengths between all pairs of nodes in the network. This matrix provides a comprehensive representation of the connectivity patterns in the network and can be used to analyze the topology and efficiency of signal transmission in the network .

In geometric networks, where nodes are physical objects and connections exist based on their physical proximity, the shortest path matrix can be particularly useful in analyzing the efficiency of signal transmission. By identifying the shortest paths between nodes, it is possible to gain insights into how signals propagate through the network and how the physical distances between nodes affect the transmission of these signals.

Overall, the shortest path matrix is an important tool in network analysis that can provide valuable information about the structure and function of networks, including geometric networks.

On the basis of the shortest path, it is possible to define other two centrality metrics:

- **closeness centrality**: It is defined as follows for the i-th node:

$$CC_i = \frac{1}{\sum_{i=j}^{n} sp_{ij}} \tag{10.1}$$

  where sp_{ij} represents the generic element of the shortest path matrix; closeness centrality measures how close a node is to all other nodes in the network. Nodes with higher closeness centrality are those that are, on average, closer to all other nodes in the network. Closeness centrality can be useful in identifying nodes that are central in terms of information flow in the network. For example, a node with high closeness centrality in a transportation network may be a hub for the efficient flow of goods or people, while a node with high closeness centrality in a social network may be a key connector between different groups of people.
- **betweenness centrality** : It is defined for each node as the number of the shortest paths passing by the node. In other words, it quantifies how often a particular node acts as a bridge or intermediary along the shortest path between pairs of other nodes in the network.

Nodes with high betweenness centrality are often considered to be critical to the network's overall connectivity and are more likely to act as bottlenecks or points of failure if they are removed from the network.

Node betweenness centrality is widely used in network analysis and can be applied to a variety of different types of networks, including social networks, transportation networks, and biological networks. It has been used in a variety of applications, such as identifying key players in a social network or predicting the spread of disease in a population.

eigenvector centrality: Also known as *eigencentrality* or *prestige score* indicates the influence of a node in the network. The eigenvector centrality for the i-th residue is defined as:

$$x_i = \sum_j A_{ij} \cdot x_j \tag{10.2}$$

where the sum extends to all nodes. This definition clearly points to a recursive computation, solved by the eigenvector decomposition of the adjacency matrix $\mathbf{A}$. According to the Perron-Frobenius theorem (*'Given a nonnegative symmetric matrix* $\mathbf{A}$*, the eigenvector* $\mathbf{v}$ *corresponding to the largest* $\mathbf{A}$ *eigenvalue has only positive components'*), the components of the eigenvector corresponding to the largest eigenvalue (called *Perron vector*) of $\mathbf{A}$ are the nodes eigenvector centrality; the fact they are positive are due to the fact $\mathbf{A}$ is nonnegative and symmetric.

In network analysis, the eigenvector centrality of a node is proportional to the sum of the centrality scores of its neighboring nodes, and the importance of a node is reflected in the centrality scores of its neighbors. This means that nodes that are connected to other highly central nodes will have higher centrality scores themselves, and this effect is amplified by the eigenvector calculation.

In biological networks, such as protein-protein interaction networks or gene regulatory networks, eigenvector centrality has been found to be particularly relevant for identifying key players or *hubs* in the network. These hubs are often critical for maintaining the integrity and function of the network, and they can have important implications for understanding disease mechanisms or designing therapeutic interventions.

For example, in a protein-protein interaction network , nodes that have high eigenvector centrality are often proteins that are involved in many different pathways or have numerous interactions with other proteins. These proteins may be particularly important for maintaining the overall structure and function of the network, and they may be potential targets for drug development.

Node degree plays also a central role in identifying network models used to represent complex systems. The distribution of the node degree is useful to define some specific network models, such as the following:

1. **Random networks**: Random networks are generated by randomly connecting nodes, with each node having a certain probability of being connected to another node. Random networks are often used as a null model for comparing against

real-world networks. In this case, the node degree distribution follows a Poisson distribution;

2. **Scale-free networks**: Scale-free networks have a few highly connected nodes (hubs) and many poorly connected nodes. They are characterized by a power-law distribution of node degrees.
3. **Small-world networks**: Small-world networks are characterized by a high clustering coefficient (nodes tend to be connected to their neighbors) and a small average path length between any two nodes.
4. **Modular networks**: Modular networks consist of subgroups of nodes (modules) that are more densely connected within the module than with nodes outside the module.
5. **Hierarchical networks**: Hierarchical networks have a nested structure, where nodes are grouped into clusters, which are then grouped into larger clusters.
6. **Bipartite networks**: Bipartite networks consist of two sets of nodes, where edges only connect nodes from different sets.

Figure 10.2 represents the node degree distribution for random (left panel), hierarchical (center), and scale-free (right panel) networks.

These models are important to predict the dynamical behavior of complex systems, since they show a typical pattern of time evolution.

The dynamical features of network models depend on the specific model used and the characteristics of the system being studied. However, in general, network models can exhibit a wide range of dynamic behaviors, including the following:

- *Synchronization*: In some network models, nodes can synchronize their behavior with their neighbors, leading to collective behavior or oscillations in the network.
- *Cascading failure*: In some network models, the failure of a single node can trigger a cascading failure that leads to the failure of many other nodes in the network.
- *Phase transitions*: In some network models, small changes in the network structure or parameters can lead to large changes in the network behavior, such as a phase transition between different matter states.
- *Resilience and robustness*: Network models can exhibit resilience and robustness to perturbations, where the network is able to maintain its function despite the removal of some nodes or edges.
- *Emergent behavior*: Network models can exhibit emergent behavior, where the behavior of the network as a whole is not simply the sum of the behavior of its individual nodes, but instead arises from the interactions between nodes.

Understanding the dynamic behavior of network models can be important for predicting and understanding the behavior of complex systems. By simulating the behavior of a network model, researchers can gain insights into how the network responds to changes in its environment or interactions with other networks, and how this affects the behavior of the system as a whole.

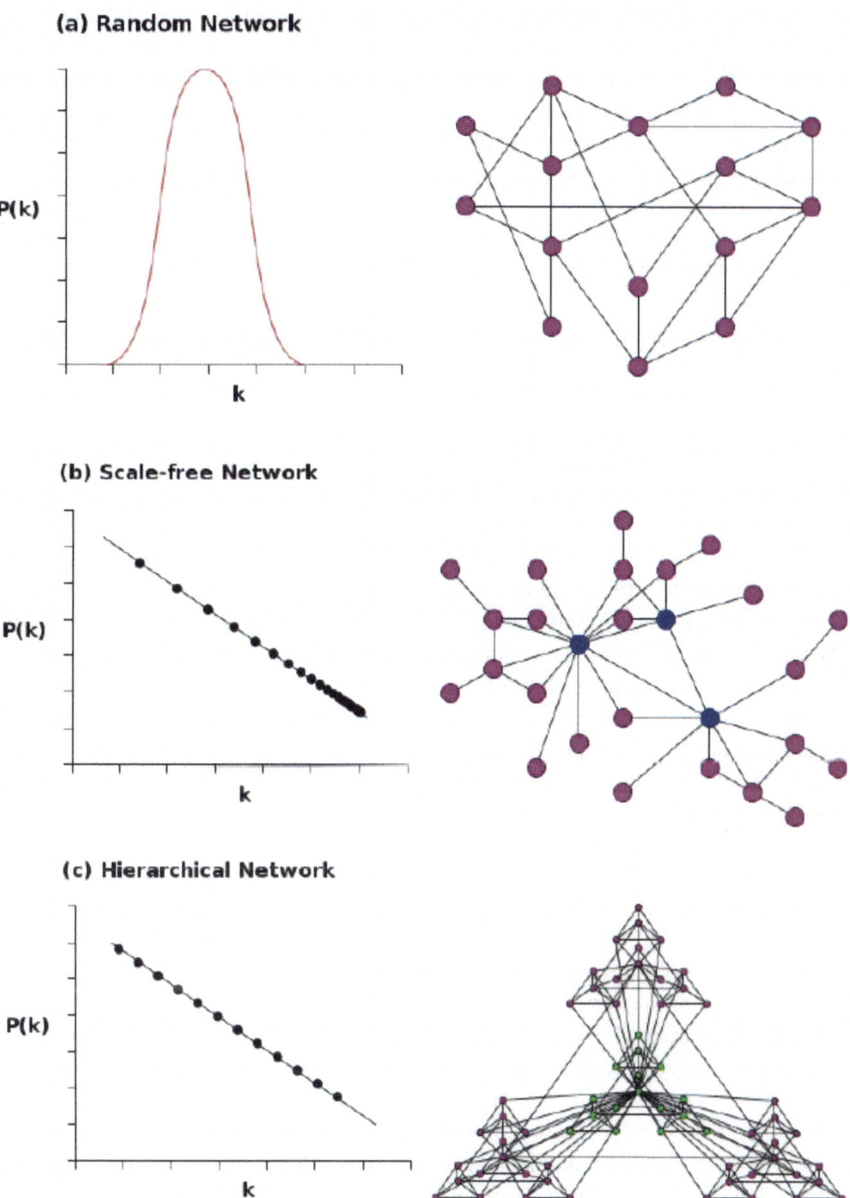

Fig. 10.2 Node degree distributions of networks: (left) random networks; center) scale-free networks; (right) hierarchical networks. Reprinted with permission from [36]

All network models are based on the quantitative description of the properties of the networks. It is necessary at this point to introduce the concept of *network topology*: it refers to the arrangement of the nodes, edges, and connections in a network and defines the structure of the network and the way in which signals flow through it. The topology of a network can have a significant impact on its performance, scalability, reliability, and security.

10.3 Systems Biology: Biological Entities as Complex Systems

The complex systems theory is a framework that helps us understand how complex phenomena, such as biological systems, emerge from the interactions of many interconnected and interdependent components. Biological systems, in particular, are highly complex systems that involve multiple levels of organization, from molecules and cells to organs and organisms.

Here are some reasons why it is important to represent biological systems through the complex systems theory:

- **Emergent properties**: One of the key features of complex systems is that they exhibit emergent properties, which are properties that cannot be predicted or explained by examining the individual components alone, but arise from the interactions among them. In biological systems, emergent properties are seen at all levels of organization, from the behavior of individual cells to the function of entire organisms. By using the complex systems theory, we can better understand how these emergent properties arise and how they contribute to the overall behavior of the system.
- **Nonlinear relationships**: Biological systems often exhibit nonlinear relationships, where small changes in one component can have large effects on the overall system. For example, a mutation in a single gene can have significant effects on the entire organism. The complex systems theory provides a framework for understanding these nonlinear relationships and how they contribute to the behavior of the system.
- **Feedback loops**: Many biological systems involve feedback loops, where the output of the system feeds back into the system as input. Feedback loops can lead to self-organization and can help the system adapt to changing conditions. The complex systems theory provides a way to model and analyze feedback loops, which can help us understand how biological systems maintain stability and adapt to changing environments.

Another important aspect is that the complex systems theory is an interdisciplinary framework integrating concepts from physics, mathematics, computer science, and other fields. By using this approach, it is possible to bring insights and tools from different disciplines to better understand biological systems. This can lead to new insights and discoveries that may not have been possible using a single disciplinary approach. We can sum up here some of the most relevant limitations and strengths of the application of the complex systems theory to the analysis and prediction of biological systems as follows:

- *Limitations*:
  1. **Modeling challenges**: Biological systems can be challenging to accurately model the behavior of all their interacting components, possibly limiting the accuracy and usefulness of predictions made using the complex systems theory.
  2. **Data limitations**: Many biological systems are still not fully understood, and there may be limited data available to model them accurately. This can limit the ability to apply complex systems theory to these systems.
  3. **Nonlinearity**: Biological systems often exhibit nonlinear relationships between their components, which can make it difficult to accurately model their behavior. Nonlinear relationships can lead to unexpected behavior and can be difficult to predict using mathematical models.

- *Strengths*:
  1. **Emergent properties**: The complex systems theory provides a lean framework to interpret emergent properties not coming directly—and simply—out of the properties of single elements in the system;
  2. **Feedback loops**: The complex systems theory is particularly useful for understanding feedback loops in biological systems, which often have a significant impact on the behavior and stability of biological systems;
  3. **Predictive power**: The complex systems theory can be used to make predictions about the behavior of biological systems. While these predictions may not always be accurate, they can provide a starting point for further experimentation and can lead to new perspectives and insights.

In summary, the complex systems theory has both limitations and strengths when applied to the analysis and prediction of biological systems. While the complexity and nonlinearity of these systems can be challenging to model, the multidisciplinary approach, understanding of emergent properties, feedback loops, and predictive power provided by the complex systems theory can lead to new insights and discoveries.

All in all, representing biological systems through the complex systems theory can help to better understand the emergent properties, nonlinear relationships, feedback loops, and interdisciplinary nature of these systems. This can lead to new insights and discoveries that may ultimately benefit human health and well-being.

Coming to the protein world, proteins are large biomolecules that play a critical role in many biological processes, including catalyzing chemical reactions, transporting molecules, and transmitting signals between cells. Proteins are composed of amino acids, which are connected by peptide bonds to form a linear chain. However, the linear sequence of amino acids does not fully explain the function and behavior of proteins, but the three-dimensional structure and the interactions between amino acids and other molecules ultimately determine the protein's function.

Protein molecules can be represented as complex networks, because they are composed of many interacting parts at different scales: the structure and, consequently, the function are determined by the intramolecular interactions between residues, and on a wider scale, they are composed of domains, motifs, and loops that interact with each other to create a complex three-dimensional structure.

Moreover, proteins interact with other proteins and molecules to perform their functions, and these interactions can also be represented as a complex system, where the structure of relationship between proteins determines the function of the whole system.

In summary, proteins can be represented through the complex systems theory, because they are composed of many interacting parts, and their function and behavior are determined by the interactions between these parts. The complex network representation of protein molecules allows to study their function and behavior in a systematic way and can provide insights into their role in biological processes and diseases.

In addition to the properties mentioned before, the hierarchy organization and self-organization properties of protein structures are also important for modeling and understanding their behavior.

Protein structures exhibit a hierarchical organization, which means they are composed of multiple levels of organization, from the primary sequence of amino acids to the tertiary and quaternary structures. Each level of organization is critical for the function and behavior of the protein. By modeling the hierarchical organization of protein structures, it is possible to gain insights into how the different levels of organization contribute to the overall behavior of the protein.

Protein structures also exhibit self-organization properties, which means they have the ability to spontaneously organize into complex structures without the need for external guidance or intervention. Self-organization properties are crucial for the function and behavior of proteins. By modeling the self-organization properties of protein structures, we can better understand how they spontaneously organize and shape complex structures, and how this determines their function.

Furthermore, self-organization is related to the concept of emergent properties, which means that the properties of a complex system are not simply the sum of its individual parts, but instead emerge from the interactions between its parts. In the case of protein structures, the self-organization of their constituent parts can lead to the emergence of new properties that are critical for protein function.

Network models are a powerful tool for analyzing and modeling complex systems, including protein networks. In the context of protein networks, a network model typically represents proteins as nodes and the interactions between them as edges.

Protein networks can be analyzed at different levels of organization, including the interactions between individual proteins and the larger protein-protein interaction network . Network models can be used to identify key proteins and pathways that are important for protein function and behavior.

One approach to building network models of protein interactions is to use experimental techniques, such as yeast two-hybrid assays, to identify pairs of interacting proteins. These interactions can then be used to construct a network model of the protein-protein interaction networks. Another approach is to use computational methods, such as homology modeling and molecular docking , to predict protein interactions and construct network models.

Once a protein network model has been constructed, network analysis techniques can be used to identify key nodes and pathways. For example, network centrality measures can be used to identify proteins that are highly connected within the network and are likely to play important roles in protein function. Pathway analysis can be used to identify groups of proteins that are functionally related and to understand the flow of information and materials through the network.

Network models of protein interactions can also be used to predict the behavior of proteins under different conditions or in response to perturbations. By simulating the behavior of the network model, it is possible to gain insights into how the network responds to changes in its environment or interactions with other molecules.

10.4 Structure-Based Protein Networks

Network models can also be used to represent the structure of proteins. In this context, a protein structure can be represented as a network of nodes and edges, where nodes represent atoms or groups of atoms in the protein structure, and edges represent the interactions between these atoms.

There are several approaches for constructing network models of protein structures. One approach is to use covalent bonds between atoms as edges in the network and represent atoms as nodes. Another approach is to use non-covalent interactions between atoms, such as hydrogen bonds or van der Waals interactions, as edges in the network.

Once a network model of a protein structure has been constructed, network analysis techniques can be used to gain insights into the structure and function of the protein. For example, centrality measures can be used to identify key atoms or groups of atoms that are important for protein stability or function. Community detection algorithms can be used to identify clusters of atoms that are functionally related, such as active sites or binding sites.

Network models of protein structures can also be used to predict the behavior of proteins under different conditions or in response to mutations. By simulating the behavior of the network model, researchers can gain insights into how changes in the structure of the protein affect its stability, function, and interactions with other molecules.

In general, the definition of a protein structural network aims at finding descriptors to address different emerging properties of the structure:

- Residues role in protein stability: residues play a different role in stabilizing the protein structure. This feature is evident when mutated forms with the same number of mutated residues show different change in stability upon mutation ($\Delta\Delta G_{unf}$);

- Quantitative global descriptors for protein stability: the computation or the experimental measurement of the ΔG_{unf} is time-burdening and expensive, while the network descriptors are often computed on the sole base of the protein structural information, so it's convenient to adopt the network paradigm to reduce the time costs for the computation of the protein stability according to its structure;
- Residues role in the protein allosteric regulation: as for the stability, the allosteric regulation is mainly addressable to a reduced number of residues, whose contribution is crucial for the regulative functional modes more in general. Identifying more active residues in this perspective can provide relevant insights also for mutation effects on protein function and more in general on the functional features of protein structures;
- Global protein functional descriptors: the identification of quantitative scores to address the global functional activity of a protein is useful to compare different protein molecular structures—such as wild type and mutants—in terms of their intramolecular connectivity (directly influencing their stability and functionality).

To sum up, structural networks are valuable tools in protein structural biology, because they provide a means of reducing the information volume, allowing for the separation of useful information from irrelevant information. This is important, because the issue of redundancy in information can impede the ability to identify relevant data when too much information is present. By reducing the amount of information to a manageable level, structural networks can help researchers identifying and analyzing critical structural features of proteins, leading to a better understanding of their function and behavior.

In addition to reducing the information volume and identifying critical structural features, structural networks are also valuable models for understanding protein dynamics and function in a faster way. These models allow researchers to analyze how individual protein residues interact with each other another, leading to a better understanding of protein folding, stability, and dynamics.

This information can then be used to predict how changes in protein structure will affect protein function, allowing, for instance, for a more efficient drug discovery and design. Furthermore, by using structural networks to analyze protein interactions, researchers can gain insights into complex biological processes such as signaling pathways and enzyme catalysis.

This approach to the representation of protein molecular structures resounds the one adopted in *chemical graphs*, representing in chemical theory the exact correspondence, where more than the specific type of atoms or the subatomic particles inside the atoms play a marginal role with respect to the topology of intramolecular bonds network. In the case of *chemical graph theory*, it has been demonstrated that this network topology is sufficient to predict molecule physical properties, such as fusion heat.

Overall, the use of structural networks represents a powerful approach for analyzing protein structure and function, allowing for faster and more efficient analysis of complex biological systems, paving the way for high-throughput screening applications, for instance, in testing a wide library of compounds as candidate drugs.

10.4.1 Protein Contact Networks

Protein Contact Networks (PCN), also known as protein residue networks, represent a valuable tool for analyzing protein structure and function. These networks are constructed by identifying the network of interactions between individual protein residues, allowing for a more detailed understanding of the complex interactions that underlie protein dynamics and function.

The forthcoming sections will present an elucidation of the principles that underlie protein contact networks/protein residue networks, encompassing their construction and potential applications in drug discovery and design. Additionally, a discussion regarding the advantages and limitations of these models will be provided in comparison to alternative network-based approaches, such as coevolutionary analysis.

Specifically, the methods used to construct protein contact networks/protein residue networks will be discussed and shown, including statistical models and machine learning algorithms. The various types of information that can be extracted from these models will be illustrated, such as residue centrality, community structure, and protein stability.

Furthermore, the relevance of these models in studying protein dynamics and function will be shown and demonstrated, particularly in relation to enzyme catalysis and protein-ligand interactions. It will be discussed how these models can be used to identify key structural features in proteins, which can then be targeted for drug discovery.

Overall, this section will provide a comprehensive overview of the principles and applications of protein contact networks/protein residue networks, highlighting their potential as powerful tools for understanding protein structure and function in a variety of biological contexts.

10.4.1.1 Contact Network Definition

The protein contact networks aim at depicting the network of the significant intramolecular interactions between residues, which stabilize the protein structure. The rearrangement of this network in protein dynamics is the key for the protein structure adaptability to environment cues.

The interactions between residues in turn depend on their mutual distances, which determine the interactions strength. Due to the fact that covalent bonds in the protein structure (peptide and disulfide bonds) range between 3 and 4 Å, the cutoff distance to determine relevant inter-residue interactions lies in a variable range around Å; in this way, only non-covalent relevant interactions are accounted for, which are liable to rearrange in protein dynamics (the covalent bonds are, on the

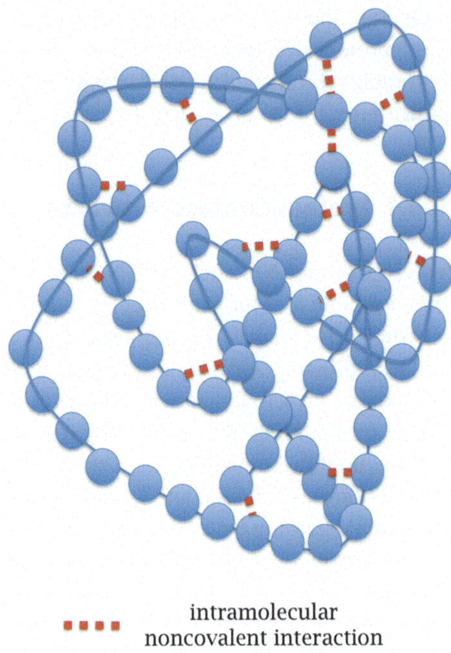

Fig. 10.3 Protein molecular structure can be represented as a tangled pearl string, where the pearls are the protein residues. The red dashed segments represent the active links between residues, notwithstanding with their distance in sequence—along the pearl string. Reprinted with permission from [23]

other hand, very stable and not varying in the typical fluctuations of the environment conditions).

In other words, the protein molecular structures can be represented as a tangled pearl string (see Fig. 10.3), where the active links exist between residues in close contact (red dashed segments in Fig. 10.3).

Protein contact networks can be represented either as unweighted or weighted graphs . In the case of unweighted graphs, the inter-residue interactions are computed on the basis of the distance between the key residue atoms (such as the α-carbons) or between the residue center-of-mass. If this distance between the i-th and the j-th residues d_{ij} lies in the cutoff range, there is a link.

Figure 10.4 represents the translation of the protein molecular structure into the corresponding distance matrix : panel a reports the three-dimensional structure of human hemoglobin, different colors depict different chains, while panel b reports the corresponding distance matrix as a heat map, hot colors correspond to large distances, where cold colors correspond to small distances.

In the unweighted graph, the mathematical representation is the *adjacency matrix* A, whose generic element is:

$$A = \begin{cases} 1 & \text{if the } i\text{-th and the } j\text{-th residues share a link} \\ 0 & \text{otherwise} \end{cases} \quad (10.3)$$

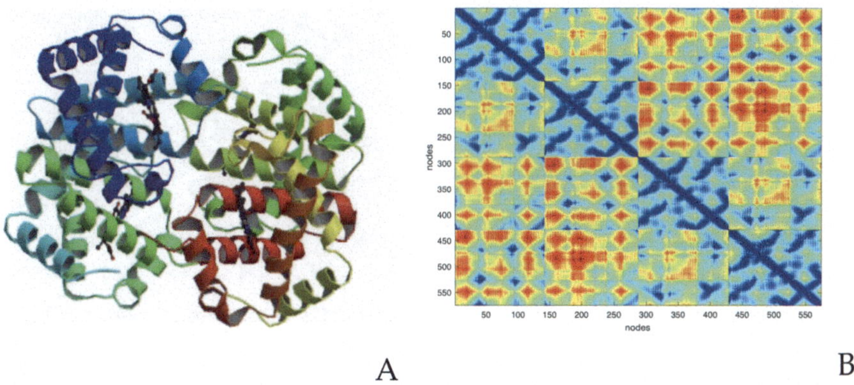

Fig. 10.4 Protein molecular structure and distance matrix representation. (**a**) human hemoglobin three-dimensional structure; (**b**) hemoglobin distance matrix

In the case of weighted graphs, the existing links are identified, s similarly to the case of unweighted graphs, if the inter-residue distance falls in a cutoff range; these links are annotated with a weight; for instance, the weight of the link between the i-th and the j-th residues w_{ij} can be the inverse of the distance d_{ij} or the number of atoms of the two residues distant less than 5 Å.

In the case of weighted graphs, the adjacency matrix is defined as:

$$A = \begin{cases} w_{ij} & \text{if the } i\text{-th and the } j\text{-th residues share a link} \\ 0 & \text{otherwise} \end{cases} \tag{10.4}$$

Figure 10.5 reports the graphical representation of the adjacency matrices for the hemoglobin protein contact network: in panel a, the black dots correspond to the active links between residues, while panel b reports the heat map of the weighted graph adjacency matrix, where the weight $w_{ij} = \frac{1}{d_{ij}}$.

10.4.1.2 Contact Network Descriptors

As previously stated in Sect. 10.2.1, on the basis of the adjacency matrix, it is possible to define several network descriptors, which correspond to specific features of the single nodes or of the whole structure. Sometimes, the descriptors at the level of the single node are named as *centralities*, highlighting the relevance of their use to address the role of single nodes (residues) in the whole network topology and dynamics.

Node degree k_i is a direct description of residues role in local stability; node degree distribution defines the type of network model. Protein contact networks escape a well-defined definition, being roughly described as small-world networks.

Figure 10.6 depicts the node degree distribution for the protein contact network model for the human hemoglobin structure (PDB code 2HHB). In panel a, the heat maps of the node degree projected on the ribbon structure represent in hotter colors

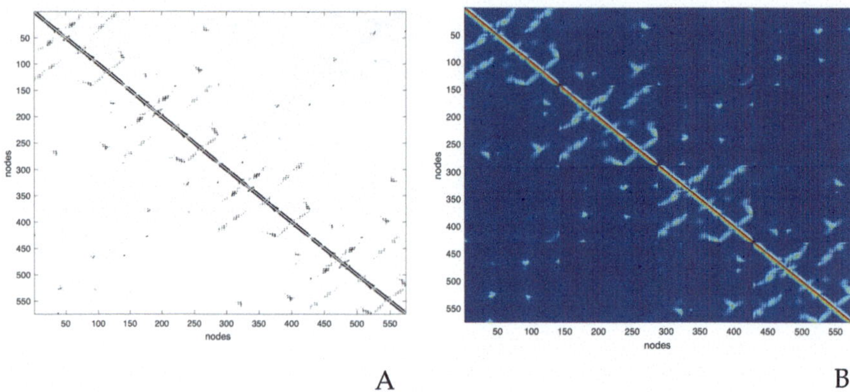

Fig. 10.5 Adjacency matrix of protein contact networks. (**a**) for an unweighted graph; (**b**) for a weighted graph, matrix points represent the inverse of the corresponding residue-residue distance in heat map

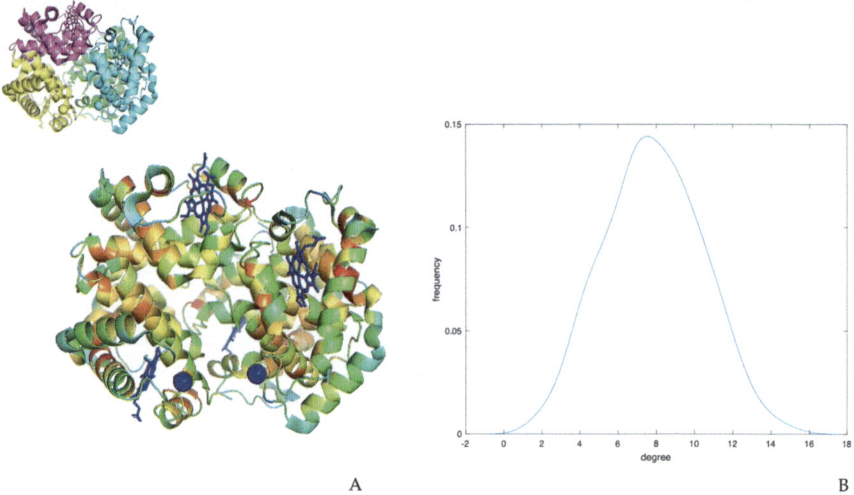

Fig. 10.6 Protein contact network (unweighted) degree for human hemoglobin (PDB code 2HHB): (**a**) node degree heat maps on the ribbon structure; (**b**) node degree distribution. For comparison, on the upper left side of the figure is represented the structure with chains highlighted in different colors)

the regions where the density of contacts is higher. The node degree distribution (panel b) quite good resembles a Gaussian centered on a mean value of 8; thus, we can say the protein contact network for the human hemoglobin follows quite well a random model.

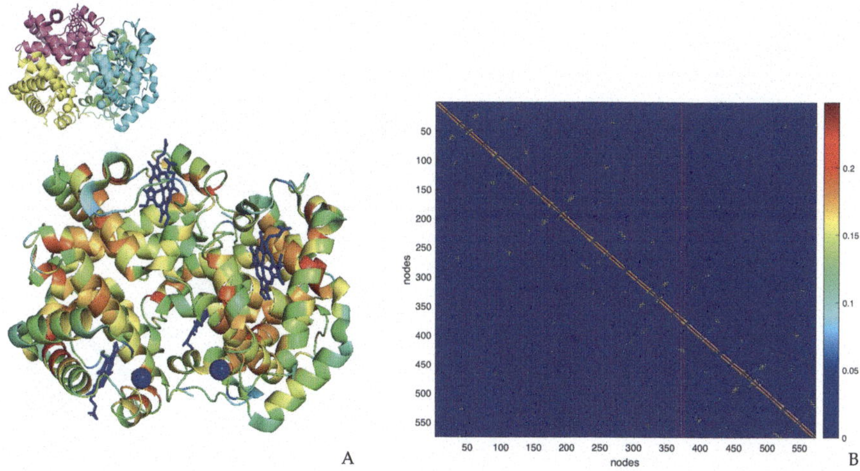

Fig. 10.7 Weighted protein contact network degree for human hemoglobin (PDB code 2HHB), the weight is the inverse of the distances between pairs of residues: (**a**) weighted node degree heat maps on the ribbon structure; (**b**) weighted adjacency matrix. For comparison, on the upper left side of the figure is represented the structure with chains highlighted in different colors)

Figure 10.7 reports the node degree heat map for human hemoglobin in the case of the weighted network (the weight is given by the inverse of the distances between residue pairs).

The average shortest path , that means the average of shortest paths between all residue pairs over the overall number of pair $\frac{N \cdot (N-1)}{2}$, also known as *network characteristic length*, is a descriptor of the signal transmission character over the whole protein molecule, being a general score of the allosteric nature of the protein molecule's functionality.

The heat map in panel a strongly resembles that in Fig. 10.6, such as the adjacency matrix of the weighted network (panel b), resembles that of the unweighted one (panel a in Fig. 10.5.

Betweenness centrality is a protein contact network descriptor that measures the importance of a residue in mediating communication between other residues in the network. Specifically, it measures the number of shortest paths that pass through a given residue.

Residues with high betweenness centrality are important for maintaining the overall connectivity and communication within the protein structure; they are often located at important structural elements, such as protein interfaces, and are involved in critical protein-protein interactions. In addition, residues with high betweenness centrality are often found in the core of the protein structure, where they play a critical role in stabilizing the overall structure of the protein.

Figure 10.8 reports the heat maps of the betweenness centrality (panel a): it is evident that residues with high values of betweenness centrality (hottest colors) are few spots in the structure, central residues in the signal transmission throughout

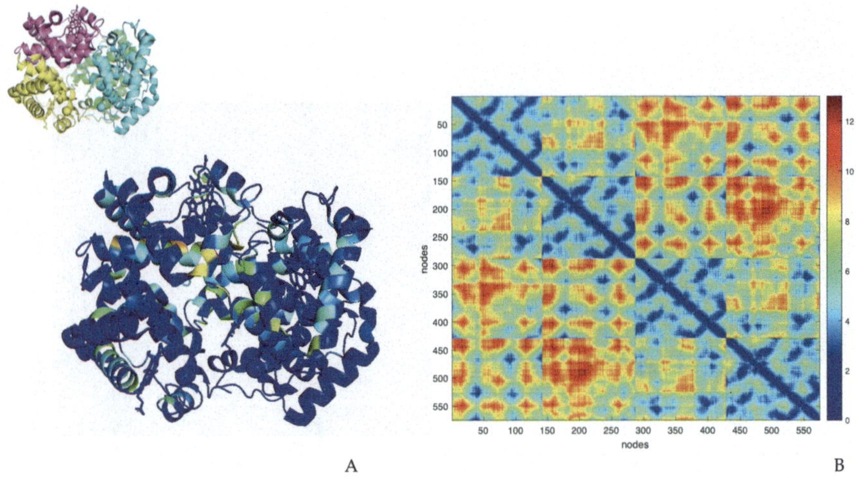

Fig. 10.8 The betweenness centrality for human hemoglobin protein contact network (PDB code 2HHB): (**a**) betweenness centrality heat maps on the ribbon structure; (**b**) shortest path matrix for the protein contact network. For comparison, on the upper left side of the figure is represented the structure with chains highlighted in different colors)

the protein molecule. Panel b of the same figure represents the shortest path: along the diagonal, the rectangles in cold colors point to the strong connectivity between residues pertaining to the same chains.

Eigenvector centrality is a measure used to identify the most important nodes in a network. In the context of protein contact networks, it can be used to identify the most influential residues in a protein structure based on their network connectivity. Specifically, eigenvector centrality can be used to calculate the importance of each residue based on how connected it is to other residues in the network. Residues with higher eigenvector centrality are considered more influential, as they have more connections to other residues in the protein structure.

Figure 10.9 reports the heat map of the eigenvector centrality for the protein contact network of the human hemoglobin. It is noticeable the high asymmetric distribution of this descriptor (high values, hottest colors in the red circle in the Figure), which in turn shows its strong dynamic nature.

The relevance of eigenvector centrality in protein functionality lies in the fact that highly connected residues are often crucial for maintaining the stability and function of a protein. In particular, allosteric regulation in folding and function is known to involve changes in the connectivity between different residues in a protein.

For example, allosteric regulation can involve changes in the interactions between residues that are located far apart in the protein structure. Eigenvector centrality can help identify such residues, as they are likely to have a high degree of connectivity with other residues in the protein structure.

In addition, eigenvector centrality can be used to predict the effects of mutations on protein function. Mutations that disrupt highly connected residues are more likely

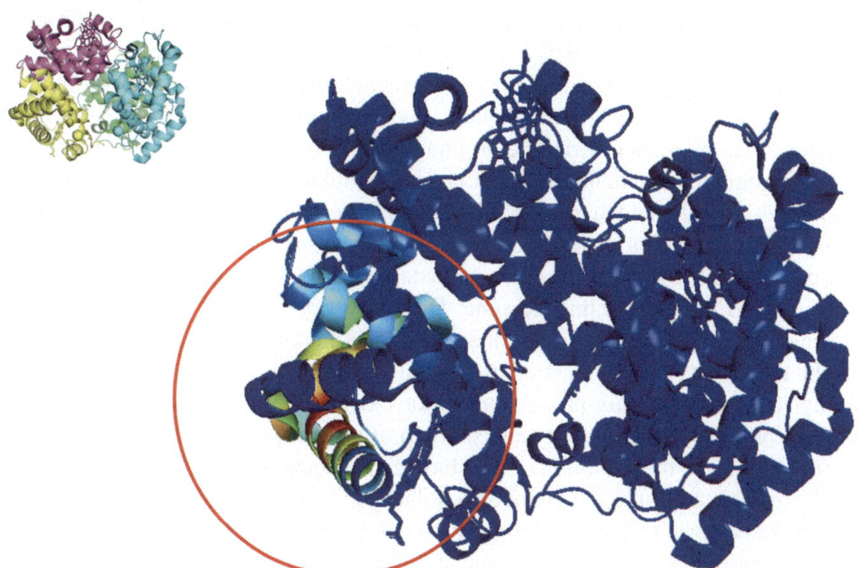

Fig. 10.9 The eigenvector centrality centrality for human hemoglobin protein contact network (PDB code 2HHB): the distribution is far from being symmetric, the high values—hottest colors, red circle—are concentrated in the yellow chain (see for comparison on the upper left side of the figure the structure with chains highlighted in different colors)

to have a significant impact on protein stability and function than mutations that disrupt less connected residues.

10.4.1.3 Network Clustering and Protein Modularity

An important feature of protein structure with a large outcome in their functionality is the *modularity*. Protein structure modularity refers to the fact that a protein can be divided into discrete, functional units or modules (chains or domains if the protein has a single chain) that can interact with other modules or with other proteins to carry out specific biological functions. Modularity is a fundamental feature of protein structure and is critical to the ability of proteins to perform a wide range of functions in living organisms.

Protein modularity is a fundamental concept in protein evolution, because it allows proteins to evolve new functions by combining existing modules in new ways. The modularity of protein structure provides a foundation for the evolution of new protein functions and the adaptation of proteins to new biological contexts.

The modularity of protein structure arises from the fact that proteins are composed of discrete structural and functional domains that can be combined and recombined in different ways to create new proteins with novel functions. This modularity allows proteins to evolve rapidly by recombining existing domains rather than having to create entirely new proteins from scratch.

One way in which protein modularity contributes to protein evolution is through gene duplication events. When a gene is duplicated, one of the copies can accumulate mutations that change the protein structure and function. By recombining domains from the original protein in new ways, the duplicated gene can evolve new functions that are distinct from the original gene. This process of gene duplication and recombination is a major driver of protein evolution.

Another way in which protein modularity contributes to protein evolution is through the acquisition of new domains from other proteins. Proteins can acquire new domains through horizontal gene transfer or through the incorporation of viral or bacterial DNA into the host genome. These new domains can be combined with existing domains to create novel proteins with new functions.

The modularity of protein structure also allows proteins to evolve new functions through changes in protein-protein interactions. By modifying the binding interfaces between proteins, the specificity and strength of protein-protein interactions can be altered, leading to the evolution of new biological functions.

The relevance of protein structure modularity to protein function is that it allows proteins to perform multiple functions by combining different modules in different ways. This modularity can be observed at different levels of protein structure, from the individual amino acid residues to larger structural domains and even entire protein complexes.

At the level of individual residues, modularity can be observed in the way that individual amino acid side chains contribute to different functional roles within the protein structure. For example, residues in the active site of an enzyme may be highly specialized for catalyzing a specific reaction, while other residues in the protein structure may be more generalist in their functional roles, contributing to overall stability or flexibility of the protein structure.

At the level of structural domains, modularity can be observed in the way that different domains within a protein structure contribute to different functional roles. For example, a kinase domain may phosphorylate other proteins, while a DNA-binding domain may recognize and bind to specific DNA sequences.

At the level of protein complexes, modularity can be observed in the way that individual proteins interact with each other to form functional complexes. For example, a transcription factor may interact with other proteins in the cell to form a complex that regulates gene expression.

The identification of modules in protein structure is a central task in computational biology. There are different computational methods to identify modules in protein structures, including clustering, network analysis, and machine learning.

The adoption of the network paradigm allows to improve the identification of modules in protein structure by applying methods of clustering to the network . In general, clustering involves grouping elements of a set on the base of similarity with respect a given properties. In the case of network clustering, the guiding scores for clustering are network descriptors, so the final result is grouping nodes according to similar values of a given descriptor. Clustering can be performed using various algorithms, such as spectral, k-means or hierarchical clustering .

As follows, it will be presented a detailed description of the spectral clustering algorithm as applied to protein contact networks. The focus is strictly on unweighted undirected graphs. *Spectral clustering* is a widely used method used for clustering data points into groups based on their similarity or distance. It relies on the eigenvalues and eigenvectors of the Laplacian matrix of the data to perform the clustering.

The spectral clustering algorithm has several advantages over traditional clustering algorithms, including the ability to handle nonlinearly separable data, the ability to cluster data with complex structures, and the ability to handle large datasets.

One drawback of spectral clustering is that it can be computationally expensive, especially for large datasets. Additionally, the choice of the number of clusters to use can be challenging and generally requires some prior knowledge (i.e., as explained in the following for the protein contact networks, the number of clusters can match the number of chains in a multimeric protein structure, and the clustering can identify specific interactions and communication patterns between chains).

Spectral clustering starts from the spectral decomposition of the Laplacian matrix **L**, defined as follows:

$$\mathbf{L} = \mathbf{D} - \mathbf{A} \tag{10.5}$$

where **A** is the adjacency matrix and **D** the degree matrix , i.e., a diagonal matrix where the diagonal entries correspond to the degree of each node. Upon the eigenvalue decomposition of the Laplacian **L**, the eigenvector corresponding to the second minor of eigenvalue v_2 (*Fiedler vector*) is selected for the clustering partition. One way to partition nodes into two clusters is by using the sign of the corresponding components of vector v_2. Specifically, nodes can be assigned to one of the two clusters based on whether the corresponding component of v_2 is positive or negative.

Spectral clustering well applies for *binary hierarchical clustering*, where the nodes are recursively partitioned into two clusters at each level of the hierarchy. As a result, spectral clustering can be particularly effective for clustering data with a hierarchical structure that can be captured by a binary tree-like organization. By using the Fiedler vector to identify natural partitions within the data, spectral clustering can help to reveal underlying patterns and structure that may be difficult to discern using other clustering techniques.

In the context of protein contact networks, spectral clustering can be a useful tool for identifying modules or clusters within protein contact networks, which are groups of highly interconnected nodes that are more densely connected to each other than to nodes outside the cluster. These modules can correspond to functional units or pathways within the protein, and understanding their organization can provide insights into the protein's overall structure and function.

By partitioning nodes into groups based on the sign of the Fiedler vector, spectral clustering can effectively identify these modules in a hierarchical manner, revealing the nested structure of the network. This can help to identify key regions of the protein that are involved in specific functions or interactions and can also aid in the prediction of new interactions or potential drug targets. In this way, spectral clustering can be a valuable tool for understanding the complex structure and function of protein contact networks.

If the number of clusters is not a power of two, such as when clustering a trimer into three clusters to emphasize the communication between chains, the process is again based on v_2. The component values of v_2 are then partitioned into the same number of clusters; in this case, the resulting clustering is no longer hierarchical.

After clustering, it is possible to compute two descriptors for single nodes:

- **participation coefficient**: this descriptor scores the role of each node in the communication between clusters. The participation coefficient P_i for the i-th node (residue) is defined as:

$$P_i = 1 - \left(\frac{k_{si}}{k_i}\right) \tag{10.6}$$

  where k_i is the node degree, while k_{si} is the node degree but only within its own s-th cluster. P_i is between 0 and 1 and is very close to unity for nodes (residues) sharing several links with nodes belonging to clusters different from the ones they belong to.

  Residues with high values of P_i are likely responsible for communication between different clusters, which can match protein chains or domains; in other words, high P residues are in charge for the allosteric signal transmission, and mutations affecting them can have a dramatic effect on the allosteric nature of the whole protein molecule. The high P residues often have also high betweenness centrality scores that confirm their central role in the allosteric regulation of the whole molecular structure.
- **z-score of the intracluster connectivity**: this descriptor plays a dual role with respect to the participation coefficient P; it is defined for the i-th node (residue) as:

$$z_i = \frac{k_i - \overline{k}_s}{\sigma_s} \tag{10.7}$$

  where $\overline{k}_s$ is the average node degree computed exclusively in the cluster to which the node belongs, where the standard deviation σ_s is also computed.

The P-z maps provide a *general cartography for complex networks*, according to *Guimerà-Amaral*; they provided a classification of nodes with respect their role within and outside the module (the equivalent of "cluster") they belong, according to the P-z space, as shown in Table 10.1.

Table 10.1 Guimerà-Amaral cartography

	Regions	z_i	P_i
Module non hubs	R_1: ultraperipheral nodes	<2.5	<0.05
	R_2: peripheral nodes		0.05–0.625
	R_3: non-hub connectors		0.625–0.8
	R_4: non-hub kinless nodes		≥0.8
Module hubs	R_5: provincial hubs	≥2.5	<0.3
	R_6: connector hubs		0.3–0.75
	R_7: kinless hubs		≥0.75

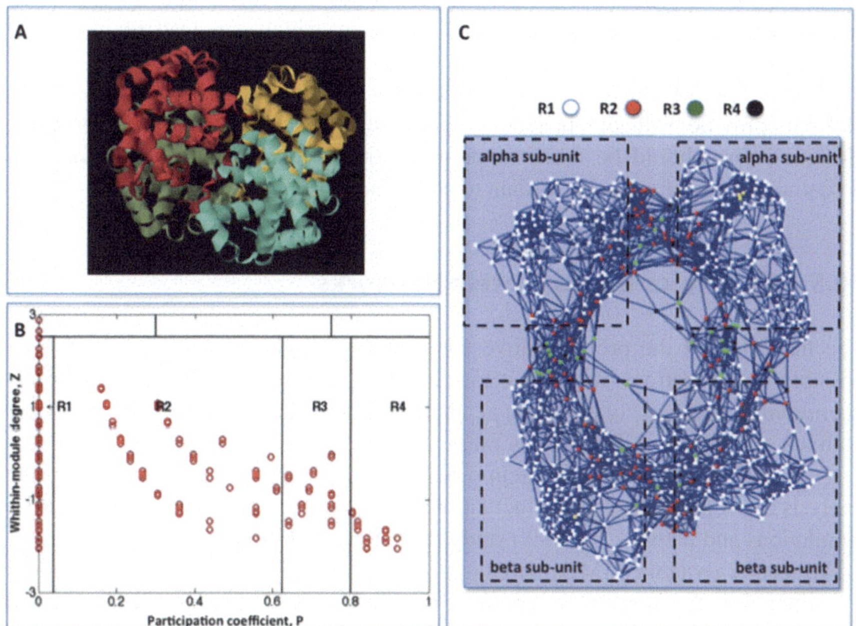

Fig. 10.10 Guimerà-Amaral cartography for hemoglobin. (**a**) Human hemoglobin molecular structure (PDB code 2HHB); (**b**) P-z map and Guimerà-Amaral cartography according to spectral clustering; (**c**) hemoglobin protein contact network: nodes colors correspond to their role in the Guimerà-Amaral cartography. Reprint with permission from [22]

Figure 10.10 reports the cartography applied to the human hemoglobin protein contact network. Panel b reports the P-z map on which the Guimerà-Amaral cartography applies. Some major features emerge as for protein contact networks:

- most of the nodes fall in the R1–R5 classes, no R6 or R7 are found (and very few in the R5 class);
- P-z maps for folded proteins show a characteristic shape, called "dentist's chair," endowed with different contour lines.

All in all, spectral clustering is a powerful tool that has found applications in the analysis of protein contact networks. It can provide a quite easy way to identify functional domains in protein molecules or complexes, providing valuable insights into the structure and function of proteins and their interactions.

Additionally, spectral clustering can handle large datasets and high-dimensional data. This is particularly important for protein contact networks, which can be very large and complex. Spectral clustering can efficiently handle such large datasets and identify patterns that might be difficult to detect using other approaches.

However, there are also limitations to the spectral clustering approach. One limitation is that it requires the choice of certain parameters, such as the number of clusters, which can affect the results. Additionally, the results of spectral clustering can be sensitive to noise in the data or outliers, which can lead to inaccurate clustering.

Overall, spectral clustering is a useful tool for analyzing protein contact networks and can provide valuable insights into the structure and function of proteins. However, care should be taken to properly choose parameters and account for potential sources of noise in the data to ensure accurate and reliable results.

10.5 Dynamics-Based Protein Networks

The flexibility of the protein native conformation produces a large ensemble of possible conformations very close to the native ones, recognized as *equilibrium fluctuations*. These fluctuations are generally small in magnitude (few Å) and occur within a subnanosecond frequency range.

Dynamics-based network protein modeling is a computational approach for the analysis of protein equilibrium fluctuations based on the molecular dynamics (MD) simulations and network theory to study the behavior of proteins.

Molecular dynamics (MD) simulations are meant to capture protein dynamics by simulating the motion of atoms within the protein structure. The classical approach to the analysis of equilibrium fluctuations is the *Normal Mode Analysis (NMA)* using all-atom empirical potentials developed for proteins. However, this approach combining MD and all-atom NMA is heavily computationally burdening, practically preventing the application to large protein molecules or complexes.

On the other hand, the application of coarse-grained (CG) models and simplified force fields allows a good description of the vibrational modes of the protein structures, with a special focus on collective motions of large macromolecular complexes.

A good simplification level is given in the dynamic network modeling of protein structures, where single residues are seen as single "beads" or point-like particles, connected to each other by elastic springs. When a force field is applied to the network, it is possible to calculate the energy associated with the interactions between nodes.

Once the network model is constructed, network analysis methods can be used to gain insights into the dynamical behavior of the protein through the computation of network descriptors, such as centrality measures, community structure, and the degree of correlation between different nodes. These analyses can reveal important information about the protein's functional mechanisms, stability, and allosteric regulation.

Dynamics-based network protein modeling provides a powerful tool for studying the behavior of proteins in a dynamic and networked context. This approach can help to uncover new insights into protein structure-function relationships as well as facilitate the design of new protein-based drugs and materials. As follows, three different models will be presented: elastic network models (ENM), anisotropic network modeling (ANM) , and energy flow networks (EFN), which present several advantages over other techniques, including computational efficiency, simplified representation, incorporation of collective motion, flexibility and versatility, and extensive validation and comparison to other methods.

10.5.1 Elastic Network Modeling

Elastic Network Modeling (ENM) is a widely used technique for studying the dynamics of proteins and other biomolecules. It is based on modeling the protein structure as a system of point particles connected by elastic springs; the protein dynamics produces as result a *harmonic potential energy function*, which describes the elastic deformation of a protein structure in response to thermal fluctuations. ENM is a simplified representation of the protein structure that focuses on the connectivity between amino acid residues and ignores the detailed atomic interactions.

The **Gaussian Network Model (GNM)** represents the protein as a network of C_α connected by springs of uniform and constant force γ if they lie within a given cutoff r_c. The term "Gaussian" refers to the fact the residues undergo Gaussian fluctuations around their mean position (*equilibrium position*); in this case, the interaction potential for a protein with N residues is:

$$V_{GNM} = -\frac{\gamma}{2} \cdot \left[\sum_{i=1}^{N-1} \sum_{j=i+1}^{N} \left(R_{ij} - R_{ij}^{0}\right) \cdot \left(R_{ij} - R_{ij}^{0}\right) \Gamma_{ij} \right] d \tag{10.8}$$

where R_{ij}^{0} and R_{ij} are the equilibrium and the instantaneous distance between residues i and j, respectively, Γ is the $N \times N$ Kirchhoff matrix, whose generic element is defined as:

$$\Gamma = \begin{cases} -1 & i \neq j,\ R_{ij} \leq r_c \\ 0 & i \neq j,\ R_{ij} > r_c \\ -\sum_{i,i\neq j} \Gamma_{ij} & i = j \end{cases} \tag{10.9}$$

Normal modes are extracted by eigenvalue decomposition of the *Hessian matrix*, which is the $3N \times 3N$ matrix whose generic element is defined on the basis of the second derivatives of the potential energy of the protein with respect to its Cartesian coordinates, as follows:

$$H_{ij} = \begin{bmatrix} \frac{\partial^2 V}{\partial X_i \partial X_j} & \frac{\partial^2 V}{\partial X_i \partial Y_j} & \frac{\partial^2 V}{\partial X_i \partial Z_j} \\ \frac{\partial^2 V}{\partial Y_i \partial X_j} & \frac{\partial^2 V}{\partial Y_i \partial Y_j} & \frac{\partial^2 V}{\partial Y i \partial X_j} \\ \frac{\partial^2 V}{\partial Z_i \partial X_j} & \frac{\partial^2 V}{\partial Z_i \partial Y j} & \frac{\partial^2 V}{\partial Z_i \partial Z_j} \end{bmatrix} \quad (10.10)$$

where X_i, Y_i and Z_i are the Cartesian coordinates of the i-th residue, while V is the interaction potential energy of the system (protein molecule).

By diagonalizing the Hessian matrix, one can obtain a set of normal modes, which are the collective motions of the protein that involve the smallest amount of energy. Hinge sites for a protein are defined by GNM at the fluctuation minima of the lowest modes (low-frequency modes). These hinge sites help maintain collective behaviors of proteins and significantly contribute to the allosteric mechanism. Figure 10.11 shows the identification of hinge sites for two conformations (tense T and relaxed R) of the human hemoglobin. It is well shown how the minima of

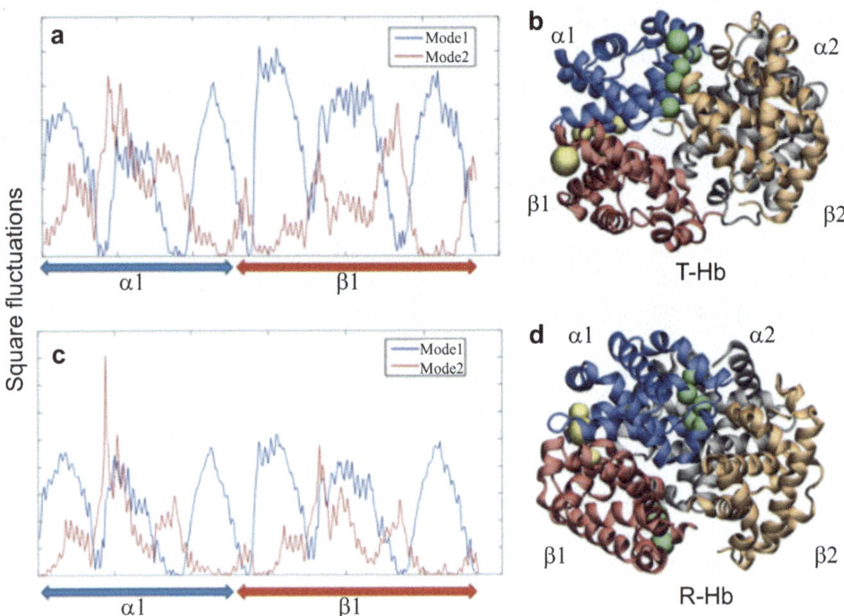

Fig. 10.11 GNM application to identify hinge sites for two conformations of human hemoglobin. (**a**) The minima of square fluctuations, based on the lowest GNM modes, predict hinges located in the interfaces between chains in T-Hb; (**b**) green and yellow beads denote hinges predicted by the first and the second GNM modes, respectively. Similar results are found for R-Hb in panels **c** and **d**. Reprinted with permission from [27]

square fluctuations pick the hinge sites, located at the interfaces between the chains in both conformations.

The GNM provides information on the magnitude of fluctuations, which are implicitly assumed to be *isotropic*, strongly reducing the size of the ensemble of independent modes ($N-1$ rather than $3N-6$ in the case of anisotropic fluctuations).

While in GNMs, the low-frequency modes identify collective motions in protein dynamics through the simplification of isotropic fluctuations around the native conformations, the **Anisotropic Network Model (ANM)** is focused on the investigation of global motions in protein dynamics, also reaching a deeper understanding of molecular mechanisms such as allosteric couplings. The basic difference with GNM is that ANM adopts an anisotropic model for fluctuations; in this case, the interaction potential for a protein of N residues is defined as:

$$V_{ANM} = \frac{\gamma}{2} \cdot \sum_{i,j}^{N} \left(\left| R_{ij} \right| - \left| R_{ij}^{0} \right| \right)^2 \tag{10.11}$$

Again, the motion in ANM is guided by the Hessian matrix $\mathbf{H}$, defined as in Eq. 10.10, which holds information about the strength and the orientation of the interactions between residues in the structure, but not on their nature. Similarly to the GNM, the counterpart of the Kirchoff matrix Γ is given in the case of ANM is $\left(\frac{1}{\gamma}\right)\mathbf{H}$; its eigenvalue decomposition results into $3N - 6$ eigenvalue and corresponding eigenvectors, reporting frequencies and shapes of the individual modes; the main difference between ANM and GNM is that GNM only provides information about the mean squared displacements and cross-correlations between fluctuations, despite the orientation of the residue-residue interactions; on the other hand, ANM includes also information about the orientation of the residue-residue interactions, providing a complete three-dimensional description of the $3N - 6$ intramolecular modes.

Finally, to compare ANM modes, it is possible to compute the similarity between the two modes $\mathbf{u_k}$ and $\mathbf{v_l}$ for proteins with two different conformations computed as inner product of their eigenvectors:

$$O(u_k, v_l) = u_k \tilde{n} v_l \tag{10.12}$$

According to the definition in Eq. 10.12, it is possible to build up maps to compare the similarity between different protein conformations upon a given number of the slowest ANM modes. Figure 10.12 shows the results of the ANM application to compare the human hemoglobin tense (T-Hb) and relaxed (R-Hb) conformations. Panel a shows the overlap maps: considering that one corresponds to perfect overlap, it is possible to infer that there is a very good compliance between the 5-th modes of the two conformations, between the 3-rd of the R-Hb and the 4-th of the T-Hb, between the 2-nd of the R-Hb and the first of T-Hb and the first of R-Hb and the

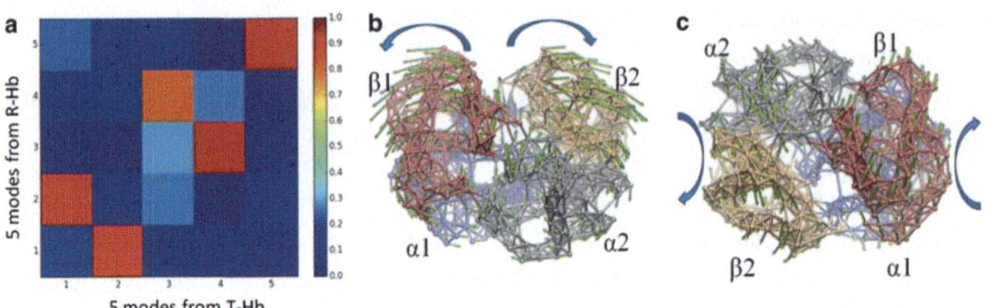

Fig. 10.12 Comparison of global motions of T and the R conformations of human hemoglobin (denoted as T-Hb and R-Hb, respectively) through ANM: (**a**) Overlaps between the five slowest ANM modes of T- and R-Hbs; (**b**) the tong-like motion, corresponding to the first mode of T-Hb or the second mode of R-Hb; (**c**) the hinge-binding rotation, corresponding to the second mode of T-Hb or the first mode of R-Hb. Reprinted with permission from [27]

2-nd of the T-Hb (all scoring 0.9); quite fair is also the overlap between 4-th of the R-Hb and the 3-th of the T-Hb (scoring around 0.75).

Panel b and c show also specific motions corresponding to normal modes among the five slowest (panel b shows the tong-link motion of the first mode of T-Hb and the 2-nd mode of R-Hb, while panel c shows the hinge binding rotation corresponding to the first mode of R-Hb and the 2-nd mode of T-Hb).

In conclusion, elastic network modeling (ENM) is a powerful tool that can provide insights into the dynamic behavior of complex biological systems. While there are different versions of ENM such as Gaussian Network Models (GNMs) and Anisotropic Network Models (ANMs), they all share the same underlying principle of modeling the interatomic interactions within a protein structure as a network of interconnected nodes and springs.

ENM has found numerous applications in the fields of drug design, protein engineering, and biomolecular simulations. Its ability to accurately capture the collective motions of a protein and its ability to predict the effects of mutations make it an indispensable tool for understanding the behavior of biological macromolecules. As computational power continues to improve, ENM is expected to become even more relevant in the years to come.

Here as follows a brief list of web servers and softwares providing services for ENM computation:

- **ProDy** is a Python-based computational toolkit, including a wide range of tools for analyzing and visualizing ENMs, ANMs, and other related models. ProDy allows users to perform normal mode analysis, elastic network analysis, and other analyses on protein structures. It also includes tools for comparing and clustering protein structures based on their dynamics. ProDy is not a web server but rather a software package that can be downloaded and installed on a local computer. It is freely available and open source and can be used by researchers with programming skills in Python.

- **elNémo** is a web interface for computing the low frequency normal modes of macromolecules through ENM. It provides different algorithms for generating ENMs and also allows the use of pre-calculated ENMs from its database. It also offers a range of analysis tools to interpret the results.
- **Web PSN** is a free web server, where through combining protein structure network (PSN) and elastic network model-normal mode analysis (ENM-NMA), it is possible to investigate the allosteric character of biological systems (proteins, nucleic acids).

Given the growing interest in elastic network modeling (ENM) and its potential to provide insights into the dynamics of complex biological systems, it is reasonable to expect that the availability and quality of web-based applications for ENM will continue to increase.

As computational power and techniques for analyzing protein structures improve, more and more researchers are likely to turn to ENM as a tool for understanding the behavior of biological macromolecules. This demand is likely to drive the development of new web-based applications and the refinement of existing ones, making ENM more accessible to a wider range of researchers and accelerating progress in this field.

10.5.2 Energy Flow Networks

Energy transport networks in proteins are responsible for the transmission of energy within a protein structure, which is essential for maintaining protein stability and function. Understanding these energy transport networks is not only important for elucidating the mechanisms underlying protein function but also for unraveling the complex process of protein folding.

Protein folding is a highly complex and dynamic process that is governed by a delicate balance between enthalpy and entropy. Energy transport networks play a critical role in balancing these opposing forces during protein folding, which ultimately determines the final folded structure of the protein.

Studying the energy flows in proteins is therefore essential for understanding the protein folding process, which has important implications for the development of drugs and therapies that target protein misfolding diseases, such as Alzheimer's and Parkinson's. Moreover, a deeper understanding of energy transport networks in proteins may also provide insights into the design of new biomolecules with unique functions, such as artificial enzymes or biosensors.

Altogether, energy flows play a significant role in allosteric regulation, as they instruct conformation changes that allow for the binding of regulatory molecules. Ultimately, the energy flows possibly outline the pathways along which the allosteric transitions occur.

Experimental methods for studying energy transport pathways have known outstanding progresses in the last years, such as the following:

1. **Fluorescence Resonance Energy Transfer (FRET)** is a technique to measure energy transfer between two fluorophores, one acting as a donor and the other as an acceptor. This method can be used to study energy transfer in proteins by labeling specific amino acids with fluorophores and measuring the transfer of energy between them.
2. **Nuclear Magnetic Resonance (NMR) spectroscopy** is a technique used to study the structural and dynamic properties of proteins. NMR can be used to measure energy transfer between atoms in a protein, providing information on energy flows within the protein.
3. **Time-Resolved Fluorescence Spectroscopy** measures the time-dependent changes in fluorescence intensity of a sample after excitation with a short pulse of light. This method can be used to study energy transfer between fluorophores in a protein and the rate at which energy is transferred.
4. **Single-Molecule Spectroscopy** allows the study of energy transfer in individual protein molecules. This method can provide information on the dynamics of energy transfer and the heterogeneity of energy transfer pathways within a population of protein molecules.

The computational approaches to detect energy transport properties in proteins are all based on molecular dynamics (MD); recently, coarse-graining methods have gained momentum, exploiting analogies to thermal transport in nanoscale materials. The application of network models allows to identify the energy transport pathways, which possibly are the energy transport channels where the allosteric signal transmission occurs.

The existence of preferential energy transport channels is linked to the non-isotropic energy flows throughout protein molecular structures, which is embedded in the folded protein geometry. This anisotropic nature of the protein dynamics is well described in terms of *local* thermal properties, such as local conductivity.

The energy transport in protein molecular structures can be also seen as a relaxation phenomenon, where a perturbation relaxes across the molecular structure through energy transport channels. Energy transport channels can be classified based on the scale they cover, as follows:

- *Long-range channels* governing the signal transmission between remote protein regions and are the basis for the allosteric regulation;
- *Short-range channels* driving the short-range extinction of local signals and responsible for the intra-domain signaling in protein structures.

In terms of which energy is transferred, the energy transport channels in protein structures can be electronic or vibrational. *Electronic energy transport* involves the movement of electrons within the molecule, and it occurs on a very fast timescale (picoseconds to femtoseconds). When a local perturbation (such as an excited electron or a photon) is introduced into the molecule, it can quickly be transferred from one electron to another in a process known as *electron hopping*. This allows the perturbation to move rapidly through the molecule, with each electron acting as a relay point.

Vibrational energy transport, on the other hand, involves the movement of energy through the molecular vibrations of the protein. This type of energy transport occurs on a slower timescale (nanoseconds to microseconds), and it involves the transfer of energy between adjacent amino acids within the protein structure. When a local perturbation is introduced into the molecule, it can excite the vibrations of nearby amino acids, which then propagate through the protein structure.

The vibrational energy is transported in a subdiffusive way, revealing a nature of the protein structure very similar to fractal objects, such as a *percolation cluster*[1] at threshold, where relatively few energy transport channels lead the anisotropic energy flows through the protein structure. In a percolation cluster, energy may be trapped and rattle around dead ends dangling from the transport channels, so statistically the spread of energy exhibits anomalous subdiffusion, which is observed in simulations on proteins and, as with a percolation cluster, can be connected to a protein's geometry and vibrational modes.

Coming to the thermodynamic nature of energy transport across a protein molecule, it can involve both enthalpic and entropic contributions, but it depends on the specific mechanism of energy transfer and the properties of the protein.

More in general, when a protein undergoes a conformation change, such as when it binds to a ligand, this can involve both enthalpic and entropic contributions: the enthalpic contribution comes from the formation or breaking of chemical bonds, while the entropic contribution comes from the change in the degrees of freedom and the increased disorder of the system.

Cooperativity can also play an important role in the energy transport channel of a protein by facilitating efficient and coordinated energy transfer between different sites or domains within the protein, enhancing the efficiency of energy transfer. For example, in a protein with multiple chromophores that can absorb and transfer energy, cooperative interactions between the chromophores can promote energy transfer between them. This can result in faster and more efficient energy transfer compared to if the chromophores were acting independently.

The classical approach to the energy transport analysis relies on creating an initial excitation in different protein regions and following the channels where energy flows.

After identifying the networks where energy transport occurs, the energy flow can be modeled by means of *Master Equations* simulations.

The Master Equation is a mathematical framework used to describe the time evolution of the probability distribution of a stochastic system, such as in the case of simulations, where Master Equation simulations allow to study of the dynamics of complex systems, such as proteins.

[1] A *percolation cluster* in a fractal solid is a network of sites which are connected to the least extent such as a long-range connectivity appears.

In Master Equation simulations, the system is modeled as a set of discrete states, and the transition probabilities between these states are described by a set of *rate constants*. By solving the Master Equation, it is possible to calculate the time-dependent probabilities of the system being in each state, as well as the time evolution of other observables of interest, such as energy or momentum.

One advantage of Master Equation simulations is that they can be used to study systems with a large number of degrees of freedom, which are difficult to simulate using other methods. However, Master Equation simulations can also be computationally expensive, especially for systems with a large number of states.

The Master Equation's rate constants can be derived either by using local energy diffusion coefficients or by fitting the energy flow results from all-atom simulations to the Master Equation.

One approach to studying energy transport in proteins is through nonequilibrium molecular dynamics simulations. In these simulations, the protein is subjected to an external force or gradient, such as a temperature or concentration gradient, and the resulting flow of energy through the protein is observed.

In order to locate the energy transport networks, it is necessary to extract the *communication maps*, based on the definition of local energy diffusion coefficients. The computation is based on the *harmonic approximation*, approaching the computation of thermal conductivity in terms of mode diffusivity or frequency-dependent energy diffusion coefficient, calculated in terms of the generic elements of the heat current operator. The mode diffusivity can be used as a score of the communication in a given mode. The intersection of large local diffusivities between residues reveals the transport channels in communication maps. This method does not require a local excitation and provides a frequency-resolved global mapping of local thermal diffusivities based on the harmonic approximation.

It is possible to define the local thermal conductivity starting from the inter-residue atom energy flow between the i-th and j-th atoms in the k-th trajectory defined as:

$$J^k_{i \leftarrow j}(t) = \frac{1}{2} \left(\mathbf{v}_i \cdot \mathbf{F}_{ij} - \mathbf{v}_j \cdot \mathbf{F}_{ji} \right) \tag{10.13}$$

where $J^k_{i \leftarrow j}$ is the inter-residue atom energy flow, mainly vibrational, between atoms i and j for trajectory k, $\mathbf{v}$ is the velocity, and $\mathbf{F}$ is the force exchanged between the two atoms (notice that it comes with a directionality).

The corresponding *energy flow* between residues A and B in the k trajectory is defined as:

$$J^k_{A \leftarrow B}(t) = \sum_{j \in A}^{N_A} \sum_{j \in B}^{N_B} J^k_{i \leftarrow j}(t) \tag{10.14}$$

where N_A and N_B are the number of atoms in residues A and B, respectively.

The corresponding *energy current* L_{AB} in the k-th trajectory is defined as:

$$L^k_{AB} = \lim_{\tau \to 0} \int_o^{\tau} \left\langle J^k_{A \leftarrow B}(t_0) J^k_{A \leftarrow B}(t + t_0) \right\rangle dt \tag{10.15}$$

The *energy conductivity* is defined accordingly as:

$$G_{AB} = (RT) \cdot L_{AB} \tag{10.16}$$

for a residue with excess energy, G_{AB} represents the net energy transferred, multiplied by (RT), across a nonbonded contact per unit time.

In other words, for a residue with an energy excess, G_{AB} represents the transferred net energy, multiplied by (RT), across a nonbonded contact per unit time. Over a large number of simulations, it is possible to build up the *energy exchange network* (EEN), whose generic element is the value of the energy conductivity for each residue-residue pair averaged for all trajectories.

Figure 10.13 shows the results of application of the EEN method to the β_2 Adrenergic Receptor, a G-protein-coupled receptor in charge for several physiological and pathological processes, such as smooth muscle relaxation, asthma, and cardiac diseases. The figure compares results for the inactive (panel a) and active (panel b)

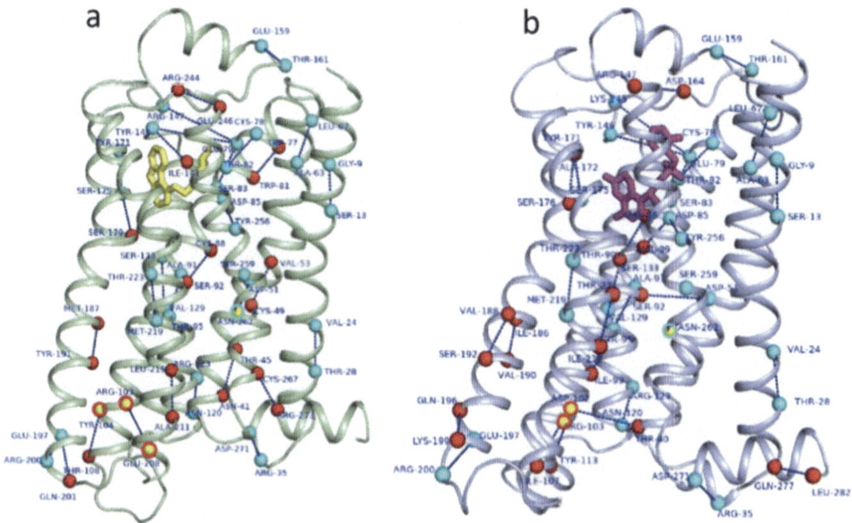

Fig. 10.13 Residues with high energy conductivity G_{AB} in EENs for the inactive (panel **a**) and active (panel **b**) state of the β_2 Adrenergic Receptor. The dotted lines connect the residue-residue pairs. Cyan residues are in contact in both states, red residues are those unique for one state, yellow residues are those located in the motif regions, circled in red or cyan if they are unique to a state or not. Reprinted with permission from [25]

states. The protein contains seven transmembrane helices, connected to each other either by either an extracellular (ECL) or intracellular (ICL) loop.

The computational implementation of the EEN method, yet being very efficient and descriptive, is not widely used, as other methods (such as ANM). However, some general tools include the computation of EEN, such as **CURP** (CURrent calculation for Proteins), a web server providing tools for the calculation of different molecular properties of protein structures, including the energy flows between residues, starting from MD simulations.

All in all, energy exchange networks are an important concept in understanding the dynamics of protein molecules. These networks represent the interconnected pathways of energy transfer that occur within a protein structure, which ultimately determine the protein's stability and functionality.

The study of energy exchange networks in proteins has important implications for a wide range of fields, from biotechnology and drug design to structural biology and biochemistry. By gaining a better understanding of the complex energy transfer pathways that occur within proteins, we can unlock new insights into the fundamental mechanisms of life and disease.

Additional Resources and Recommended Literature

General primers of systems biology are given by [10, 20]. A general review on Protein Contact Networks can be found in [24]; a general sketch for Energy Flows in proteins can be found in [32], such as a perspective in Elastic Network Modeling of protein dynamics is in [21].

Computational Protein Binding 11

The meeting of two personalities is like the contact of two chemical substances: if there is any reaction, both are transformed.

C. Jung

Abstract

Why it is important to know this material?

The computational approach to protein binding has become a crucial methodology in drug design and discovery, with a special focus on the virtual screening of drug candidates.

What is the key idea?

The key idea is to use the molecular docking and dynamics simulations to evaluate the affinity of proteins with ligands and biomacromolecules.

What is necessary to know already?

It is required to have a solid background in the protein binding mechanisms and thermodynamics and in molecular dynamics simulations (theory and applications).

11.1 Introduction

In the intricate world of biology, a multitude of essential processes, from cellular signaling to enzymatic reactions, are orchestrated by the complex interactions between proteins and their environment. Understanding the mechanisms underlying protein-ligand binding, allostery and protein-protein interactions has been a longstanding goal in molecular biology. These interactions are pivotal for modulating cellular function, regulating pathways, and executing biological responses.

L. Di Paola, *Fundamentals of Molecular Bioengineering*,
https://doi.org/10.1007/978-3-031-42022-1_11

This chapter delves into the realm of computational methods dedicated to unraveling the intricacies of protein binding, allostery, and protein-protein interactions. By harnessing the power of computational approaches, it is possible to explore the dynamic interplay between proteins and their environment, providing valuable insights into the driving forces and structural determinants governing these interactions.

Protein binding entails the formation of transient or stable complexes between proteins and other molecules, such as small ligands, DNA, RNA, or other proteins. Elucidating the binding process and characterizing the binding affinity, kinetics, and specificity are key objectives in studying protein-ligand interactions. Computational methods, including molecular docking, virtual screening, and free energy calculations, play a pivotal role in predicting and analyzing these interactions, facilitating the design of novel therapeutics and understanding biological processes.

Allostery, on the other hand, represents a fascinating phenomenon wherein a protein's function is modulated by a distant site through long-range communication. Allosteric regulation provides an additional layer of complexity and regulation to protein function. Computational methods, such as molecular dynamics simulations, network analysis, and energy landscape approaches, enable to decipher the intricate allosteric networks and understand the underlying mechanisms of allostery.

Protein-protein interactions form the foundation of many biological processes, governing cellular signaling, enzymatic reactions, and assembly of macromolecular complexes. Understanding the intricate details of these interactions is essential for deciphering cellular pathways and designing therapeutics targeting protein-protein interfaces. Computational methods, including protein docking, molecular dynamics simulations, and coevolutionary analysis, aid in exploring the vast conformation space and energetics involved in protein-protein interactions.

Through computational methods, researchers can leverage the vast amounts of experimental data, structural information, and theoretical models to gain deeper insights into the molecular basis of protein binding, allostery, and protein-protein interactions. These methods provide a powerful toolkit for unraveling the underlying principles, driving forces, and dynamics that govern the delicate dance of proteins within the complex *milieu* of biological systems. In this chapter, we will delve into these computational approaches, highlighting their strengths, limitations, and applications in deciphering the complexities of protein interactions and their functional implications.

11.2 Computational Prediction of Protein-Ligand Binding Sites

Computational prediction of binding sites poses a significant challenge in computational biology, aimed at identifying regions in proteins likely to interact with molecules like drugs or small compounds. This information plays a central role in drug design and in the full understanding of the drug mechanisms. Several methods exist for predicting binding sites.

One approach involves leveraging known protein-ligand complexes to identify amino acids involved in binding interactions. Alternatively, machine learning algorithms can predict binding sites using protein amino acid sequences. These algorithms are trained on datasets of known binding sites and applied to new proteins for predictions.

The field of computational binding site prediction is rapidly advancing, with continuous method development and improved accuracy, due to the growing computational capacities and availability. Today, the computational identification of binding site is an essential tool in drug discovery and to reveal the molecular drug mechanisms. The prediction accuracy, on the other hand, requires continuous improvement to provide more and more reliable predictions.

The challenges to face to improve the methods of computational prediction of binding sites can be grouped as the following:

- *Availability of training data*: Accurate machine learning algorithms rely on ample training data. However, the limited number of known binding sites poses a challenge in training accurate models.
- *Diversity of binding sites*: Binding sites for different ligands exhibit significant diversity, complicating the development of methods capable of predicting binding sites for a wide range of ligands.
- *Complexity of protein structure*: Protein structure influences binding properties, making it challenging to develop accurate prediction methods, particularly for proteins with complex structures.

Despite these challenges, computational binding site prediction holds promise and has the potential to revolutionize drug discovery. With continued improvement in accuracy, these methods will become indispensable tools for drug discovery and unraveling drug mechanisms.

More in general, the fields where the protein binding and the identification of binding sites play a key role are the following:

1. *Drug Discovery*: Determining protein binding is crucial for identifying and developing new drugs. Understanding the interactions between proteins and potential drug molecules helps in designing compounds with optimal binding affinity and specificity.
2. *Enzyme Substrate/Inhibitor Interactions*: Protein binding studies aid in characterizing enzyme-substrate and enzyme-inhibitor interactions. Determining binding affinities and kinetics helps elucidate enzymatic mechanisms, optimize enzyme activity, and design inhibitors for therapeutic interventions.
3. *Receptor-Ligand Interactions*: Receptors play a critical role in cellular signaling. Experimental determination of receptor-ligand interactions provides insights into signaling pathways, drug targeting, and the development of therapeutics for various diseases.
4. *Protein-DNA Interactions*: Protein binding to DNA regulates gene expression and other DNA-dependent processes. Understanding protein-DNA interactions

helps unravel transcriptional regulation, DNA repair mechanisms, and genome organization.

5. *Protein-RNA Interactions*: RNA-binding proteins control RNA processing, stability, and translation. Experimental determination of protein-RNA interactions sheds light on posttranscriptional regulation, RNA-protein complexes, and RNA-based therapeutics.
6. *Protein-Lipid Interactions*: Proteins interact with lipids to form membrane structures and influence membrane-related processes. Investigating protein-lipid interactions provides insights into membrane protein function, lipid-protein signaling, and membrane dynamics.
7. *Antibody-Antigen Interactions*: Understanding the binding between antibodies and antigens is crucial in immunology, vaccine development, and diagnostics. Experimental determination of antibody-antigen interactions aids in identifying epitopes, characterizing immune responses, and designing targeted therapies.
8. *Protein Engineering and Design*: Protein binding studies guide the engineering and design of proteins with improved binding properties. By studying binding interactions, it is possible to modify proteins for enhanced affinity, specificity, stability, or altered ligand recognition.
9. *Structural Biology*: Determining protein binding provides critical information for structural biology studies. The structural characterization of protein complexes and their binding sites aids in understanding protein function, molecular recognition, and the development of structure-based drug design strategies.

These are just a few examples of the broad range of applications in protein science where experimental determination of protein binding is highly relevant. Accurate and detailed knowledge of protein binding interactions is pivotal for advancing our understanding of biological processes and has significant implications in various fields, including medicine, biotechnology, and bioengineering.

11.2.1 Experimental Methods and Theoretical Modeling of Protein Binding

The identification of protein-ligand binding sites is a crucial stage to understand molecular mechanisms involving protein systems. Experimental methods for protein-ligand binding can be classified based on the techniques used and the information they provide. Here are some common classification categories for protein-ligand binding methods:

- *Biophysical Techniques*: These techniques are based on the changes in biophysical properties upon binding, providing quantitative data on binding affinity, kinetics, thermodynamics, and structural changes:
  - Surface Plasmon Resonance (SPR)
  - Isothermal Titration Calorimetry (ITC)

  - Fluorescence-based methods (Fluorescence Polarization, Fluorescence Resonance Energy Transfer, etc.)
  - Microscale Thermophoresis (MST)
  - Differential Scanning Fluorimetry (DSF)
- *Structural Techniques*: These methods elucidate the three-dimensional structures of protein-ligand complexes, providing atomic-level details of binding interactions and supporting the structure-guided drug design and mechanistic understanding.
  - X-ray Crystallography
  - Nuclear Magnetic Resonance (NMR) Spectroscopy
  - Cryo-Electron Microscopy (Cryo-EM)
- *Mass Spectrometry-based Binding Assays* (e.g., Surface Acoustic Wave Mass Spectrometry, Native Mass Spectrometry): These techniques work by ionizing and analyzing the mass-to-charge ratio (m/z) of protein-ligand complexes, providing information about binding stoichiometry, interactions, and structural changes, contributing to the understanding of protein-ligand binding dynamics and drug discovery.
- *Thermodynamic Techniques*: These techniques focus on the experimental determination of the thermodynamic properties to quantify the binding affinity:
  - Equilibrium Dialysis
  - Fluorescence-based Thermal Shift Assays
  - Circular Dichroism (CD) Spectroscopy
- *Kinetic Techniques*: These methods measure the rates of association and dissociation, providing insights into binding kinetics, on/off rates, and transient intermediates, crucial for understanding binding mechanisms, drug efficacy, and target interactions:
  - Stopped-Flow Spectroscopy
  - Rapid Kinetic Binding Assays

It is worth noting that these classification categories are not mutually exclusive, as some methods can provide multiple types of information. Researchers often employ a combination of techniques to obtain a comprehensive understanding of protein-ligand binding properties, leading to valuable insights in drug discovery, enzyme kinetics, and structure-based design.

While experimental methods provide valuable insights into protein-ligand binding, computational methods have emerged as powerful tools for predicting and modeling these interactions. By harnessing the power of computer algorithms and simulations, computational methods offer distinct advantages in terms of cost-effectiveness, time efficiency, and the ability to explore a wide range of ligands. However, it is important to understand the differences between computational and experimental methods to appreciate their unique contributions to the study of protein-ligand binding.

The theoretical approach to the quantitative description of the protein-binding process is based on the theory of the *molecular recognition*. In general, molecular recognition is the specific interaction between two or more molecules through non-covalent bonding such as hydrogen bonding, metal coordination, hydrophobic forces, van der Waals forces—interactions, halogen bonding, or resonant interaction effects. In addition to these direct interactions, solvents can play a dominant indirect role in driving molecular recognition in solution. The host and guest involved in molecular recognition exhibit *molecular complementarity*.

Molecular recognition plays an important role in biological systems and is observed in between receptor-ligand, antigen-antibody, DNA-protein, sugar-lectin, RNA-ribosome, etc.

There are a number of different theories that have been proposed to explain molecular recognition, as previously described in Chap. 7. One of the most popular theories is the *lock-and-key model*. The lock-and-key model states that the ligand and the protein have complementary shapes that fit together like a lock and key. This complementary shape allows the ligand to bind to the protein with high affinity. In this case, the stiff complementarity is based on the local shape of protein molecule in its surface, looking for the best match with the shape of the ligand.

However, the lock-and-key model does not take into account the fact that proteins can change their shape. In reality, the protein surface and the ligand can undergo a slight change in shape when they bind to each other. The *induced fit* model includes the change in shape upon ligand binding, allowing for a more specific and stronger interaction between the protein and the ligand (high affinity).

Several recent theories have been proposed to enhance the computational prediction of protein-binding sites. These theories focus on understanding the principles and mechanisms underlying protein-ligand interactions, which are crucial for accurate prediction of binding sites. Here are some of the key theories:

1. *Binding Site Conservation Theory*: This theory suggests that protein binding sites tend to be conserved across related protein structures, indicating functional importance. By analyzing the conservation patterns, computational methods can identify conserved residues and predict binding sites based on evolutionary information.
2. *Energetic Hotspots Theory*: According to this theory, protein-ligand binding is driven by specific energetic hotspots within the binding site. Computational approaches aim to identify these hotspots by analyzing the local energy contributions, such as favorable van der Waals interactions, hydrogen bonding, and electrostatic interactions. By focusing on the energetic hotspots, computational methods can predict the most likely binding regions within a protein.
3. *Shape Complementarity Theory*: This theory emphasizes the importance of geometric and shape complementarity between a protein and its ligand in determining binding affinity. Computational algorithms employ geometric matching algorithms, surface complementarity metrics, and shape descriptors to predict binding sites based on their compatibility with ligand shapes.
4. *Dynamic Flexibility Theory*: Recognizing the dynamic nature of protein-ligand interactions, this theory suggests that protein flexibility plays a crucial role in binding. Computational methods, such as MD simulations, are employed to capture the conformation changes and flexibility of both the protein and the

ligand. By considering the dynamic behavior of the system, these methods can identify potential binding sites that may be accessible only in specific conformations.

5. *Ligand-Based Prediction Approaches*: In addition to protein-centric theories, ligand-based approaches aim to predict protein-binding sites based on common structural and chemical features shared among ligands that bind to similar proteins.

Computational methods employ techniques such as ligand clustering, pharmacophore modeling, and machine learning to identify common ligand features associated with binding sites and predict potential binding regions in a protein. These recent theories, focusing on binding site conservation, energetic hotspots, shape complementarity, dynamic flexibility, and ligand-based approaches, provide a foundation for the computational prediction of protein-binding sites. By integrating these theoretical concepts into computational algorithms and models, researchers strive to improve the accuracy and efficiency of predicting protein-ligand interactions and guide drug discovery efforts.

11.2.2 Computational Approaches for Prediction of Protein-Ligand Binding Sites

The specificity of protein-ligand binding is due to specific chemical nature of the ligand and to the specific shape and chemo-physical properties of the binding site. The solution of this problem is crucial to understand to gain a deep understanding of the binding mechanism, predict affinity, and stability of the protein-ligand complex.

The computational approaches for predicting protein-ligand binding sites play a crucial role in drug discovery, protein function annotation, and understanding protein-ligand interactions. In the last years, the number of available methods and web tools for the prediction of the binding has ballooned.

Here are some common computational approaches used for predicting protein-ligand binding sites:

- **Structure-based models**, as the name suggests, rely on the three-dimensional structure of proteins to predict and analyze protein-ligand binding interactions. These models utilize information about the protein's active site or binding pocket to understand how ligands interact with the protein and to predict their binding affinity. Here are some specific structure-based applications:
  1. *Molecular Docking*: Molecular docking predicts the optimal binding conformation and affinity of a ligand within a protein's active site or binding pocket.
  2. *Structure-based Virtual Screening*: This approach involves screening large compound libraries against a protein's structure to identify potential ligands, using techniques such as shape-based matching, pharmacophore modeling, and docking-based scoring.

3. *De Novo Design*: *De novo* design generates new ligands with desired properties using computational algorithms to explore chemical space and predict their potential binding interactions with the protein.

These structure-based models provide diverse approaches for predicting protein-ligand binding sites, incorporating principles such as geometric analysis, evolutionary conservation, machine learning, and fragment-based strategies. They aid in the identification and characterization of binding sites, facilitating the exploration of protein-ligand interactions and assisting in drug discovery and design efforts.

- **Sequence-based Methods**: Sequence-based models for predicting protein-ligand binding sites leverage information derived from the protein's primary sequence. These models utilize sequence conservation, patterns, and motifs to infer potential binding sites. Here are some key aspects of sequence-based models:

1. *Homology Modeling*: Homology modeling, also known as comparative modeling, predicts the three-dimensional structure of a protein based on its sequence similarity to a known structure. By identifying a suitable template protein with a known structure, homology modeling can provide insights into the potential binding sites of the target protein based on the corresponding regions in the template.
2. *Evolutionary Conservation Analysis*: Evolutionary conservation analysis examines the degree of sequence conservation across multiple homologous proteins to identify functionally important residues, including those involved in ligand binding sites. Highly conserved residues are more likely to play a critical role in ligand recognition and binding. Methods such as phylogenetic profiling, coevolution analysis, and sequence motif analysis are used to identify conserved residues associated with binding sites.
3. *Pattern and Motif Analysis*: Sequence-based models may employ pattern and motif analysis to identify characteristic amino acid sequence patterns or motifs associated with ligand binding sites. These patterns can represent key residues involved in ligand recognition or binding interactions. Methods such as regular expressions, hidden Markov models (HMMs), or profile-based approaches can be employed to identify and analyze such sequence motifs.
4. *Binding Site Prediction from Ligand Similarity*: Sequence-based models may exploit ligand similarity information to predict potential binding sites. By analyzing the binding sites of proteins with known ligands, these models can identify conserved sequence regions or motifs that are likely to bind similar ligands. This approach assumes that proteins with similar ligands may share common binding sites or binding mechanisms.

These sequence-based models provide insights into potential ligand binding sites starting from the protein's primary sequence. They are particularly useful when experimentally determined structures or direct structural information are unavailable. By analyzing sequence conservation, patterns, and motifs, these models can guide further investigations into the binding sites and functional roles of proteins, aiding in drug discovery, functional annotation, and understanding protein-ligand interactions.

- **Ligand-based Methods**: Ligand-based models, also known as ligand-centric or ligand-driven models, focus on utilizing information derived from known ligands to predict protein-ligand binding sites. These models do not rely on the protein's structure but instead exploit the characteristics and properties of ligands to infer potential binding sites. Here are some key approaches of ligand-based models:
  1. *Pharmacophore Modeling*: Pharmacophore modeling identifies common chemical features or spatial arrangements shared among a set of active ligands. These features, such as hydrogen bond acceptors, donors, hydrophobic regions, or aromatic rings, are used to define a pharmacophore model. The model can then be used to screen databases of compounds to identify molecules that fit the pharmacophore, indicating potential binding sites.
  2. *Quantitative Structure-Activity Relationship* (QSAR): QSAR models the relationship between ligand structural features and their biological activities. QSAR models are trained on a dataset of ligands with known activity values, and various molecular descriptors are calculated to characterize the ligand structures. These models can then predict the activity of new ligands and identify potential binding sites based on their structural features.
  3. *Ligand Similarity and Molecular Fingerprints*: Ligand-based models often utilize ligand similarity methods to predict binding sites. Ligands with similar chemical structures or physicochemical properties are assumed to bind to similar regions on the protein. Ligand similarity can be assessed using molecular fingerprints, which encode structural or chemical information about ligands and enable comparisons to identify potential binding site matches.
  4. *Fragment-based Approaches*: Fragment-based models focus on analyzing smaller fragments or substructures of ligands to infer binding sites. These fragments are typically derived from known ligands or fragment libraries, with a special focus on peptide libraries. By identifying binding fragments and analyzing their locations and interactions, fragment-based approaches can provide insights into potential binding sites and guide ligand design and optimization.
  5. *Ligand Clustering and Scaffold Analysis*: Ligand-based models may employ clustering techniques to group ligands based on structural similarity. The determination of clusters of ligands that share similar chemical features is based on statistical methods and can identify potential binding sites. Scaffold analysis focuses on identifying common structural frameworks among ligands and relates them to binding sites. It helps identify conserved structural motifs associated with ligand binding.

  Ligand-based models offer valuable insights into binding sites by exploiting ligand characteristics and properties. They are particularly useful when protein structure information is limited or unavailable. By leveraging ligand information, these models can guide the identification of potential binding sites, aid in virtual screening, ligand design, and the exploration of ligand-protein interactions.

- **Hybrid Methods**:
  Integrated Approaches: These methods combine multiple computational techniques from different classes. For example, combining molecular docking with MD simulations, or integrating ligand-based and structure-based approaches, can enhance the accuracy and reliability of binding site predictions. It's important to note that these classes are not mutually exclusive, and some methods may overlap between categories. However, this classification provides a broad overview of the different computational approaches used for predicting protein-ligand binding sites.

It is important to note that combining multiple approaches and integrating various data sources often leads to improved predictions. Different methods have their strengths and limitations, and the choice of approach depends on the available data, the specific characteristics of the protein-ligand system, and the research objectives.

11.2.2.1 Computational Web Tools for the Prediction of Protein-Ligand Binding Sites

The concept of predicting protein binding sites has evolved significantly over the years, driven by advancements in computational methods and the integration of AI techniques. In the early stages, the focus was primarily on geometric analysis and the exploration of physicochemical properties of protein structures. It became possible to identify binding sites by considering factors such as solvent accessibility and surface shape complementarity. This led to the development of early web tools, like CASTp, which enabled the identification of potential binding pockets.

As computational methods progressed, more sophisticated algorithms were introduced for binding site prediction. Tools like SiteMap and Fpocket utilized geometric and physicochemical properties to identify potential binding sites. Molecular docking tools, such as AutoDock and DOCK, allowed to predict ligand binding orientations within protein structures. These methods provided valuable insights into the spatial arrangement of protein-ligand interactions.

A significant advancement in binding site prediction came with the integration of machine learning techniques. Machine learning algorithms have the ability to learn patterns and features from training data, greatly improving prediction accuracy and efficiency. Tools like P2Rank and 3DLigandSite employed machine learning algorithms to predict binding sites based on evolutionary and structural features. By leveraging known protein-ligand complexes as training data, these approaches built predictive models that could identify potential binding sites.

In recent years, the rise of deep learning and AI has further revolutionized binding site prediction. *Deep learning algorithms*, such as Convolutional Neural Networks (CNNs) and Recurrent Neural Networks (RNNs), have been applied to learn complex patterns and representations from protein structures. Tools like

DeepSite have leveraged deep learning techniques to enhance binding site prediction accuracy.

Moreover, AI methods, including reinforcement learning and generative models, are being explored to generate novel predictions of ligand binding sites. These AI-driven approaches enable the discovery of previously unknown or cryptic binding sites, opening up new avenues for drug discovery.

Another important integration has been the integration of MD simulations into binding site prediction methods. MD simulations allow researchers to capture the dynamic behavior of protein-ligand complexes, and the same protein dynamics is an additional precious information. By exploring conformation changes and flexibility, these simulations provide insights into transient and cryptic binding sites that may not be evident from static structures alone. Tools like MDpocket have integrated MD simulations to analyze the dynamic aspects of binding site identification.

The integration of AI techniques, such as machine learning and deep learning, with traditional computational methods has significantly improved the accuracy, efficiency, and scalability of binding site prediction. AI methods can handle complex features and interactions, leading to improved prediction models and the discovery of novel binding sites. This integration continues to drive innovation in the field, enabling more precise and robust prediction models for protein-ligand binding interactions.

In summary, the evolution of binding site prediction has been marked by advancements in computational methods, from geometric analysis to more sophisticated algorithms integrating machine learning, deep learning, and AI techniques. The integration of MD simulations has further enhanced our understanding of dynamic aspects of binding sites. The use of AI continues to push the boundaries of prediction accuracy and efficiency, facilitating drug discovery efforts and aiding in protein engineering endeavors. As follows, a more detailed description of few tools for each class described in Sect. 11.2.2, providing a general primer for protein-ligand binding sites prediction.

Structure-Based Methods for Prediction of Protein-Ligand Binding Sites

The structure-based methods for predicting binding sites are a general approach based on geometrical and biophysical properties, which allows a quite accurate prediction with rather limited computational burden.

Here as follows are listed the three highly regarded web tools among the structure-based methods, known for their ease of use and high accuracy in structure-based binding site prediction:

- COACH (Confidence-Oriented Accurate Combined Hierarchical) is a user-friendly web server that combines multiple methods to accurately predict binding sites. It integrates sequence, structure, and ligand information to provide comprehensive predictions. COACH offers a straightforward interface and generates detailed reports with annotations and visualizations. The tool has demonstrated high accuracy and reliability in predicting binding sites for a variety of protein structures.

- SiteMap: SiteMap is a widely used web tool for the prediction of binding sites in protein structures. It employs a combination of geometric and physicochemical properties to identify potential binding sites. SiteMap offers a user-friendly web interface with intuitive input options. The tool analyzes protein structures and generates predictions based on the energetics and shape complementarity of potential binding sites. It provides comprehensive reports with detailed annotations and visualizations of predicted binding sites. SiteMap is a currently available web tool that offers a user-friendly interface and accurate prediction of binding sites in protein structures. It leverages geometric and physicochemical properties to identify potential binding sites and provides detailed analysis to aid researchers in their investigations.
- 3DLigandSite: 3DLigandSite is a widely used web server for the prediction of ligand binding sites in protein structures. It utilizes a combination of ligand-based and structure-based approaches to identify potential binding sites. 3DLigandSite offers a user-friendly interface with easy-to-use input options. The server performs protein structure analysis and generates predictions based on the ligand-binding site similarity to known complexes. It provides detailed visualizations and annotations, aiding in the interpretation of the predicted binding sites. 3DLigandSite is a currently available web server that provides a user-friendly interface and accurate prediction of ligand binding sites in protein structures. It offers automated analysis and comprehensive predictions, assisting researchers in identifying potential binding sites for further investigation.

Sequence-Based Methods for Prediction of Protein-Ligand Binding Sites

As previously detailed, the purely sequence-based methods are focused on protein systems lacking of structural information. However, two factors are limiting more and more these applications: the growing availability of 3D protein structural information, due to the growing accuracy of structural biology methods for proteins recalcitrant to crystallization (NMR and Cryo-EM) and the availability of computer-aided methods for protein structure prediction (such as AlphaFold), which reduces the set of structures whose structural information is practically not achievable. However, some tools are available on the web, such as the following:

- HoTS: Sequence-based prediction of binding regions and drug-target interactions: HoTS is a deep learning method able to predict ligand-protein binding sites, trained on a large database of binding-sequence and ligand. HoTS can predict binding regions for a given protein sequence—ligand pair via object detection employing transformers.
- *HybridPBRpred*: HybridPBRpred is designed for the prediction of protein-binding residues, generating numeric score for each residue in the input protein sequence that quantifies putative propensity for protein interaction. Larger values of propensity denote higher likelihood to interact. HybridPBRpred combines prediction of protein-binding residues generated by SCRIBER (predictor that was trained using structures of protein-protein complexes) and DisoRDPbind (predictor that was trained using disordered protein-binding residues), with the underlying goal to produce complete set of the protein-binding residues.

In summary, sequence-based methods for predicting ligand binding sites serve a specific purpose in the research domain. They are valuable for screening large databases of ligands to identify potential binding interactions with a target protein. These methods find utility in diverse areas, including the repurposing of commercial drugs and assessing drug toxicity on a proteome-wide scale. They are particularly useful for identifying off-target binding, where a drug interacts with proteins other than its intended target, offering insights into the broader effects of drug-protein interactions at a systemic level.

Ligand-Based Methods for Prediction of Protein-Ligand Binding Sites
Ligand-based methods leverage information from known ligands to infer potential binding sites on the protein. These methods are valuable for screening large databases of compounds, repurposing drugs, and assessing off-target effects. They leverage ligand similarity, molecular fingerprints, pharmacophore modeling, and quantitative structure-activity relationship (QSAR) models.

Several web tools provide user-friendly platforms for applying ligand-based methods and predicting protein-ligand binding sites. These tools aid in drug discovery, lead optimization, and exploring ligand-protein interactions. As follows are three popular web tools that utilize ligand-based methods for predicting protein binding sites:

- *SuperPred*: SuperPred is a web server that employs ligand-based approaches to predict protein binding sites. It uses a large collection of ligand-binding site templates from the Protein Data Bank (PDB) to identify similar binding sites in a target protein. Users can input a protein structure or sequence, and SuperPred compares it against its database of ligand-bound structures to predict potential binding sites. The server provides a ranked list of predicted binding sites along with information on ligand similarity, binding site residues, and visualization tools.
- FINDSITE: FINDSITE is a web server that employs ligand-based methods for virtual screening of ligand-protein binding, targeted to drug discovery. It utilizes a library of ligand-binding site templates derived from the PDB to identify potential binding sites in a target protein. Users can submit a protein structure or sequence, and FINDSITE compares it to its template library using a scoring function that considers geometric and physicochemical complementarity. The server provides a list of predicted binding sites ranked by their scores, along with visualization tools and additional information about potential ligands.

These web tools utilize ligand-based methods to predict protein binding sites by leveraging information from ligands and their interactions. They offer user-friendly interfaces and provide valuable insights into potential binding sites, aiding in drug discovery, functional annotation, and the exploration of protein-ligand interactions.

Hybrid Methods for Prediction of Protein-Ligand Binding Sites

In the field of protein-ligand binding site prediction, hybrid computational approaches that combine multiple methods have gained prominence. These hybrid methods leverage the strengths of different techniques, such as structure-based and sequence-based approaches, to enhance the accuracy and reliability of binding site predictions. This section highlights a selection of currently available web tools that employ hybrid computational approaches for predicting protein-ligand binding sites.

- COACH-D: COACH-D is an advanced web server that integrates both structure-based and sequence-based methods for binding site prediction. It combines the capabilities of COACH (Confidence-Oriented Accurate Combined Hierarchical) and DISPLAR (DISorder-Prediction for Ligand-Binding ARchitectures). COACH-D utilizes protein sequence information, structural analysis, and disorder prediction to identify potential binding sites. The web server offers a user-friendly interface, allowing users to input protein structures or sequences for comprehensive binding site predictions.
- ConCavity: ConCavity is a web tool that integrates sequence conservation and structural analysis for protein binding site prediction. It combines information from protein structures and evolutionary properties to identify potential binding regions. The tool analyzes surface shape, solvent accessibility, and sequence conservation to generate predictions. It provides detailed reports with confidence scores and interactive 3D visualizations. Users can input protein structures to explore binding sites through its user-friendly interface. ConCavity assists in drug discovery, protein engineering, and functional characterization by identifying potential binding sites within proteins.

These two hybrid computational web tools exemplify the power of combining multiple techniques in protein-ligand binding site prediction. By integrating structure-based and sequence-based methods or structural alignment and pharmacophore-based approaches, these tools aim to provide more accurate and comprehensive predictions. Researchers can leverage these web tools to gain insights into potential protein-ligand interactions, aiding in drug discovery and design efforts.

11.3 Computational Approach to Allostery

The identification of the protein-ligand binding sites is only the first step of the description of the ligand-protein interactions. The allosteric regulation of the protein-ligand binding is the way proteins act as sensors and are the basis of many regulatory biological processes. The allosteric response of protein molecules is due to an interplay of environment stimuli with the structure and the signal entering the molecule is transmitted throughout the whole molecular structure.

Therefore, the computational approach to describe allosteric regulation is based on the identification of signaling pathways across the proteins molecular structures. The methods to describe allosteric signaling are focused on three major features of the protein molecular structures:

- *modularity*: Signal transmission occurs between distinct functional domains, which communicates through few residues with a distinct, key role in signal transmission;
- *coupled motions*: The allosteric signal transmission is based on the coupled dynamics of residues in distal regions of the protein structure;
- *energy change and transport*: The binding of ligands and effectors to protein molecules produces a change in the protein energy and stability (ΔG_{unfold}); the allosteric signal can be therefore seen as an energy transfer process and the energy flows outline the allosteric pathways as such.

Computational approaches provide valuable insights into the structural, energetic, and dynamic aspects of allosteric systems, aiding in the design of novel therapeutic strategies and the optimization of drug candidates. This wide-ranging section explores the diverse computational methods employed to describe allosterism, encompassing techniques such as molecular dynamics simulations, protein structure prediction, network analysis, and machine learning algorithms:

- *Molecular Dynamics Simulations*: Molecular dynamics (MD) simulations can provide detailed information about the conformation changes and dynamic behavior of allosteric proteins. These simulations enable exploring the free energy landscapes, identifying key residues involved in allosteric communication, and investigating the impact of ligand-binding on protein dynamics.
- *Protein Structure Prediction*: Accurate prediction of protein structures is essential for understanding allosteric mechanisms. Computational methods such as homology modeling, threading, and ab initio folding algorithms aid in predicting the three-dimensional structure of proteins, including their allosteric sites. These techniques enable the generation of models of allosteric proteins and the identification of potential allosteric sites for further investigation.
- *Network Analysis*: Allosteric regulation often involves long-range communication between distant sites within a protein. Network analysis methods, such as graph theory and statistical coupling analysis, help in elucidating the allosteric networks by mapping the residues and interactions involved in transmitting allosteric signals. These computational approaches allow the identification of critical residues and pathways responsible for allosteric communication, shedding light on the underlying mechanisms.
- *Machine Learning Algorithms*: Machine learning techniques have gained prominence in the field of allosteric modeling. These algorithms can analyze large-scale datasets, extract patterns, and make predictions regarding allosteric behavior. By leveraging machine learning, it is possible to develop predictive models for allosteric sites, classify allosteric proteins, and screen potential allosteric modulators from vast compound libraries.

The application of computational methods has revolutionized the study of allosterism, providing valuable insights into the complex mechanisms underlying this phenomenon. MD simulations, protein structure prediction, network analysis, and machine learning algorithms offer complementary approaches to understanding allosteric regulation at different levels of detail. By integrating experimental data with computational models, researchers can accelerate the discovery and design of allosteric modulators, leading to advancements in drug development and therapeutic interventions.

In the following, the focus is on network-based approaches to identify allosteric mechanism, defining the role played by residues in allosteric signal transmission.

11.3.1 Network-Based Computational Approaches for Allostery Quantitative Identification

The traditional understanding of allosteric regulation has largely been rooted in the structural and functional characterization of these macromolecules. However, in recent times, a paradigm shift toward network-based computational approaches has reshaped our perspective on allostery as a part of a complex system behavior, well described by the network paradigm.

The network-based computational methodologies to unravel and describe allostery generally fall into two main categories. The first category focuses on the identification of "modules" and describing allostery in terms of these modules interacting or "chatting" with each other. In this view, the macromolecule is divided into distinct functional units or modules, and allostery is the result of interactions between these units. The role of residues (nodes) in this inter-module communication is directly linked to their centrality in the allosteric signal transmission.

The second category centers around the identification of interaction pathways—the molecular "highways" along which the allosteric signal is transmitted. This perspective sees the macromolecule not as a collection of separate units, but as a whole, with allosteric regulation arising from the propagation of structural and energetic changes along specific pathways. These communication pathways correspond to the network shortest paths, and all network descriptors based on the shortest paths definition, such as betweenness centrality, address the role of single nodes (residues) into the allosteric signal transmission.

From a general point of view, the signal transmission in protein structures can be described in different ways on the basis of the network paradigm application to protein structure and function, such as in the following:

1. **Molecular Dynamics (MD) Simulations**: MD simulations model the physical movements of atoms and molecules over time, providing a dynamic view of the system. In the context of signal transmission, MD simulations can help visualize how a signal or perturbation at one site propagates through the protein

structure. One can capture the conformation changes and motions involved in signal propagation.
2. **Normal Mode Analysis (NMA)**: NMA is a technique that models the vibrational modes of a system. In the context of biomolecular networks, it can help identify collective motions that might contribute to signal transmission. It simplifies the system by considering only the low-frequency, large-amplitude vibrational modes which are often related to biological function, including allostery.
3. **Markov State Models (MSM)**: MSMs are mathematical models used to describe the conformation changes of biomolecules. They partition the conformation space into a set of states and model the transitions between these states as a Markov process. This can provide insights into the potential paths of signal transmission and their probabilities.
4. **Graph Theoretical Approaches**: In these models, a protein is represented as a graph where nodes correspond to residues or atoms, and edges represent interactions between them. Different metrics can be used to model signal transmission, such as shortest path (the path with the fewest edges between two nodes), betweenness centrality (how often a node lies on the shortest path between other nodes), or network flow models.
5. **Information Theoretical Approaches**: These methods, based on principles from information theory, quantify the signal transmission in terms of information flow. Mutual information is a common metric used to measure how much information is shared between two residues or sites.
6. **Elastic Network Models (ENMs)**: ENMs are simplified models that represent a protein as a network of beads connected by springs. They can provide insights into the collective motions of proteins that might be involved in signal transmission.

It's important to remember that while these methods can provide valuable insights into signal transmission, they also have their limitations and assumptions. For instance, MD simulations require considerable computational resources, while NMA and ENMs often ignore the effects of solvent and thermal fluctuations. Therefore, these methods should be used in conjunction with each other and with experimental data to get a comprehensive understanding of signal transmission in biological networks.

In this section, we will explore key network-based computational methods that fall under each of these categories. These methods include Protein Contact Networks (PCNs), Energy Exchange Networks (EEN) and Elastic Network Model (ENMs), and Anisotropic Network Model (ANMs). Eventually, the section will include the description of the application of the machine learning techniques, which thanks to their capabilities can be included into the other approaches to improve the accuracy of description and prediction.

Despite the tremendous advancements, it's also important to acknowledge the limitations of these computational methodologies and to stress the necessity of integrating them with experimental data and other computational approaches. This section aims to provide a comprehensive understanding of the network-based approaches to allosterism, illuminating the current state of the art and the path forward.

11.3.1.1 Structural Networks and Network Modularity

The network of non-covalent intramolecular interactions allows proteins to adapt to their environment. As outlined in Sect. 10.4, the Protein Contact Networks (PCNs) approach effectively captures this critical aspect of protein structure. By utilizing network descriptors, one can characterize the functional processes involved in protein interactions with their environment.

PCNs excel in capturing the essence of signal transmission through the structural connections formed by non-covalent interactions between residues. The signal transmission characteristics of a protein molecule are described by PCNs based on two key features:

- The *shortest path matrix* provides a global-level description of a protein molecule's propensity for allostery. Its average value and the presence of small values for shortest paths between residues at the active site and distal regions in the protein molecule indicate potential allosteric communication. These shortcuts traverse the molecular network wiring, facilitating efficient transmission of allosteric signals.
- *Protein modularity* refers to the allosteric nature of a protein molecule, which arises from communication between functional modules that may overlap with functional domains. The identification of node clusters strongly suggests corresponding functional modules or domains, and the signal transmission between these clusters, as described by signal transmission theory, effectively characterizes the transmission of allosteric signals between functional modules.

In addition to the average shortest path, which serves as a global descriptor for a protein molecule's allosteric characteristics, the shortest path matrix also helps identify the *betweenness centrality*. This local-level (single residue) descriptor quantifies the number of shortest paths that pass through a specific node. Residues with higher betweenness centrality play a more central role in the transmission of allosteric signals. It is worth noting that this descriptor primarily distinguishes allosteric properties related to binding regulation, especially in single-chain structures. In multichain structures, the betweenness centrality not only describes the transmission of signals relevant to binding regulation but also reflects communication within a chain, contributing to the maintenance of local stability. Figure 11.1 reports betweenness centrality heat map for the deoxyhemoglobin structure.

The betweenness centrality in the figure is higher in the regions close to the heme group. Figure reports the ratio of the distance/shortest path, referred to HYS87 (active site residue); panel a shows the heat map of the ratio over the ribbon structure (HYS87 in magenta spheres). The hottest colors highlight the allosteric spots for the residue, which are very effectively connected to it despite their distance. The same ratio is plotted in panel b (Fig. 11.2).

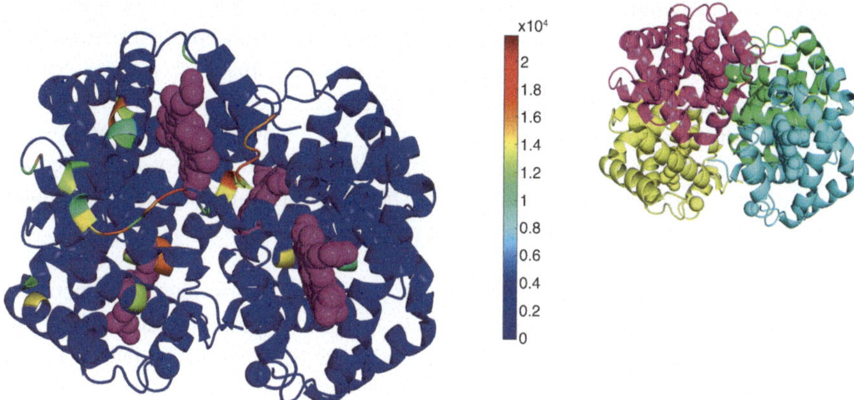

Fig. 11.1 Betweenness centrality heat map for human deoxyhemoglobin (PDB code 2HHB, left panel); in magenta spheres the heme group, the active site of hemoglobin. Residues with high betweenness centrality (hotter colors) are those close to heme groups. The right panel shows in different colors the four chains of the multimeric hemoglobin

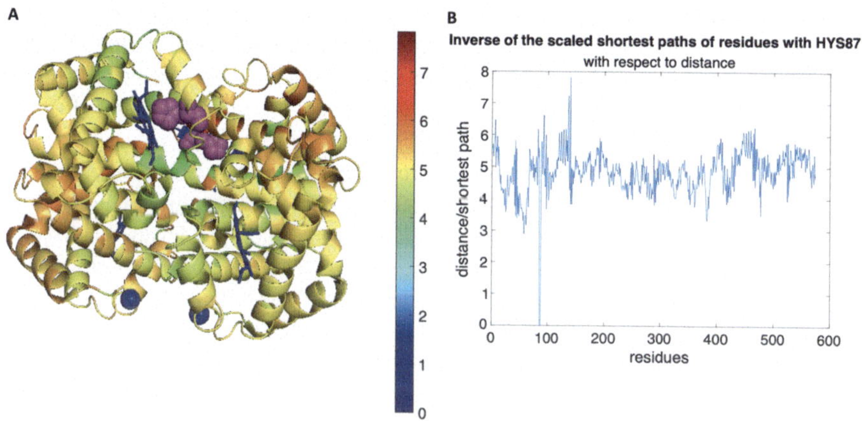

Fig. 11.2 Distance over shortest path for the HYS87, a residue in the active site (magenta spheres in panel **a**): (**a**) heat maps of the ratio distance/shortest path; (**b**) plot of the ratio distance/shortest path. The highest values represent allosteric hotspots connected in a very efficient way to HYS87, despite the large distances

Additionally, the modularity of the protein structure provides valuable insights into the global perspective of allosteric signal transmission. The hierarchical nature of protein structures, mirroring the process of folding through the formation of increasingly complex assemblies of simpler modules (such as secondary elements as building blocks for tertiary assembly), gives rise to distinct modular regions with distinct functional roles. Identifying these domains, traditionally based on recognizing folding patterns, can also be achieved through network clustering, where PCNs clusters represent groups of nodes (residues) identified by maximizing intracluster connectivity. This approach, widely employed in network functional mapping, is highly useful for identifying functional modules in protein structures.

In the field of signal transmission analysis, network clustering plays a crucial role in understanding the structure and dynamics of complex communication networks. Network clustering refers to the process of identifying groups or clusters of interconnected nodes that exhibit higher internal connectivity compared to connections with nodes outside the cluster. This concept is highly relevant in signal transmission analysis due to the following reasons:

1. *Identifying functional modules*: Network clustering helps identify functional modules within a communication network. In signal transmission analysis, these modules represent groups of nodes that work together to perform specific signal processing tasks or carry out particular functions. By identifying these modules, it is possible to reach a deeper description of the organization and coordination of signal transmission processes, leading to a deeper understanding of the underlying mechanisms.
2. *Analyzing communication efficiency*: Network clustering allows the analysis of communication efficiency within a network. In signal transmission, efficient communication is vital to ensure reliable and rapid transmission of signals. Clustering the network allows to assess how effectively information flows within and between clusters. This analysis aids in identifying potential bottlenecks, areas of high traffic, or critical nodes that play a significant role in maintaining efficient signal transmission.
3. *Unveiling hierarchical structures*: Network clustering can reveal hierarchical structures in communication networks. In signal transmission, hierarchical organization often exists, where nodes are arranged in multiple levels, each contributing to different aspects of signal processing. The application of clustering algorithms can uncover these hierarchical structures and help understanding how different levels of organization interact and contribute to signal transmission, enabling a comprehensive analysis of the network's functionality.
4. *Detecting signal propagation pathways*: Clustering provides insights into the pathways through which signals propagate within a network. Grouping nodes based on their interconnectivity can identify dominant pathways or routes that signals take while being transmitted. This information helps in optimizing signal transmission strategies, identifying potential vulnerabilities, and designing robust communication networks.
5. *Understanding network resilience*: Network clustering allows the assessment of network resilience in signal transmission. Resilience refers to a network's ability to maintain its functionality even when faced with failures or disruptions. The analysis of clusters and their interconnections allows to determine how the loss of specific nodes or clusters affects the overall signal transmission performance. This knowledge aids in designing resilient network architectures and developing strategies to mitigate potential failures or disruptions.

In parallel, in the context of allosteric regulation in proteins, network clustering techniques, especially hierarchical (spectral) clustering, play a crucial role in unraveling the mechanisms and functional implications of allosteric interactions, as follows:

1. *Identification of allosteric communication pathways*: Protein contact network clustering enables the identification of long-range communication pathways that mediate allosteric signals within the protein structure. The application of hierarchical clustering techniques to PCNs identifies clusters of residues that participate in these communication pathways, shedding light on the residues and pathways involved in allosteric signal propagation.
2. *Detection of allosteric hotspots*: Clustering of protein contact networks helps identify allosteric hotspots, which are regions of the protein that are crucial for transmitting allosteric signals. Allosteric hotspots often correspond to clusters of residues that undergo conformation changes or exhibit altered interactions upon binding of an allosteric effector. Hierarchical clustering can highlight these clusters, providing insights into the specific residues and regions that are functionally relevant for allosteric regulation.
3. *Analysis of dynamic allosteric networks*: Hierarchical clustering techniques, combined with molecular dynamics simulations, can reveal dynamic allosteric networks within protein structures. Allosteric regulation involves the modulation of protein dynamics and conformation changes. By clustering protein contact networks at different time points during simulations, it is possible to track changes in clustering patterns and identify dynamically coupled residues or regions that are involved in allosteric communication. This analysis contributes to understanding the dynamic nature of allosteric regulation and its implications for protein function.
4. *Prediction of allosteric sites and modulators*: Protein contact network clustering can aid in the prediction of allosteric sites and the identification of potential allosteric modulators. By clustering residues based on their contact patterns, it is possible to identify clusters that are enriched in residues located near known allosteric sites or in regions involved in allosteric communication. This information can guide the identification of novel allosteric sites and the design of small molecules or drugs targeting allosteric regulation.
5. *Elucidation of allosteric mechanisms*: Network clustering approaches, such as hierarchical clustering, help elucidate the underlying mechanisms of allosteric regulation. By analyzing the clusters and their interconnections within protein contact networks, it is possible to infer the pathways and mechanisms by which allosteric signals propagate and influence protein function. This understanding of allosteric mechanisms is crucial for deciphering the principles governing allosteric regulation and for designing strategies to modulate protein function through allosteric regulation.

All in all, in the realm of allosteric regulation in proteins, network clustering techniques, particularly hierarchical (spectral) clustering, provide valuable insights into the communication pathways, allosteric hotspots, dynamic networks, prediction of allosteric sites and modulators, as well as the underlying mechanisms of allosteric regulation. These clustering methods enable to analyze protein contact networks with the final aim at uncovering the residues, clusters, and pathways involved in allosteric signal propagation. This knowledge contributes to a better understanding of allosteric regulation and opens avenues for designing novel strategies to modulate protein function through allosteric means, with implications for drug discovery and therapeutic interventions.

As seen in Sect. 10.4.1.3, spectral clustering of PNCs, a hierarchical method based on the eigenvector decomposition of the Laplacian matrix describing the PCNs, allows to identify functional modules with a good accuracy. The cluster partition of PCNs identifies in different clusters distinct functional modules. These modules communicate to each other, and this communication guides the allosteric regulation. Thus the cluster identification is the first application of PCNs to provide insight in the allosteric signal transmission.

Another central task in the analysis of the allosteric character of a protein molecular structure is to address the residues role in transmitting the allosteric signal. Upon network clustering, the participation coefficient P describes the role of single residues in the communication between network clusters (see Sect. 10.4.1); this descriptor was demonstrated to reveal the allosteric regions in proteins in different ways:

1. the regions of proteins with $P > 0$ are those participating in communication between clusters; residues with values of $P \geq 0.75$ are *allosteric hotspots*, regions with a strong allosteric character;
2. computing the values of ΔP between the unbound and bound forms (before and after binding) it is possible to describe the activation/deactivation switch character of some regions, a dynamic description of structural basis of the allosteric regulation.

Figure 11.3 shows the application of the network clustering to the structure of unbound cell-wall invertase (PDB code 2AC1), a plant enzyme playing a central role in carbohydrate metabolism. Panel a reports the partition in two clusters, which well identifies the two functional domains in the single chain of cell-invertase. Panel b shows the map of the participation coefficient P of the unbound cell-wall invertase, showing in blue the central regions of the two functional domains, not interested by the transmission of allosteric signals between domains, while hotter colors in the frontier between the two domains highlight the residues responsible to transmit between the two clusters (functional domains) the allosteric signal. Panel c shows the map of ΔP between the unbound and the bound form with sucrose (PDB code 2OXB), addressing the switch region, activating(hot colors)/deactivating (cold colors) upon binding.

All in all, the application of the PCNs helps outlining the structural basis of the allosteric regulation, starting from the consideration the structural networks of protein molecules are responsible to transmit the environmental cues throughout the

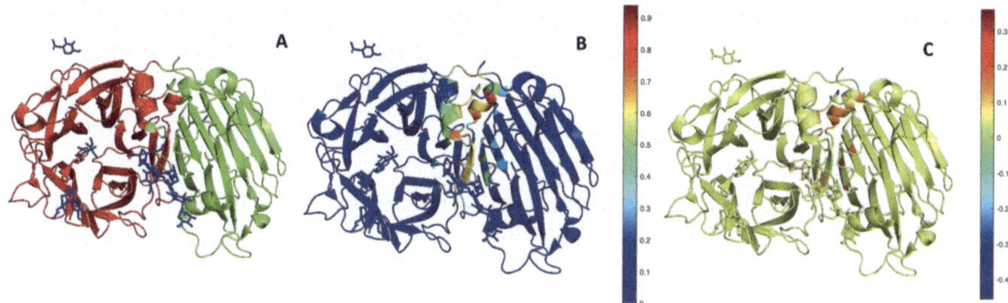

Fig. 11.3 The application of the network clustering to the cell-wall invertase: (**a**) clustering partition (2 clusters) of the unbound cell-wall invertase (PDB code 2AC1); (**b**) participation coefficient heat map of the unbound cell-wall invertase (PDB code 2AC1); (**c**) $\Delta P = P_{unbound} - P_{bound}$ map of the cell-wall invertase, the bound form is in complex with sucrose (PDB code 2OXB)

molecular structure. Overall, the application of protein contact networks to study allosteric regulation provides a structural framework for understanding the dynamic interplay between distal regions of a protein. It offers insights into the transmission of allosteric signals, identification of key residues and communication pathways, and the potential for targeted interventions. This method holds promise for advancing our knowledge of allosteric mechanisms and facilitating the development of novel therapeutics.

11.3.1.2 Elastic Network Models (ENMs) for Allostery Analysis

The coupled motions of different regions of the protein structure are at very basis of all cooperative processes, mainly, allosteric regulation.

The use of the Elastic Network Models (ENM), comprising the Gaussian Network Model (GNM) and the Anisotropic Network Model (ANM), represents a significant advancement in our understanding of the allosteric mechanism within protein structures. Allostery refers to the phenomenon where a change in one region of a protein structure triggers a response in another distal region, often leading to functional alterations. Investigating these complex and intricate mechanisms is crucial for unraveling the underlying principles governing protein behavior.

Traditionally, studying allosteric communication in proteins has been a challenging task due to the intricate nature of their three-dimensional structures and the dynamic nature of their function. As seen in Sect. 10.5.1, ENMs provide valuable computational approaches that allow researchers to probe the allosteric coupling between distal regions. These models simplify the protein structure by representing it as a network of interconnected nodes, with each node representing a residue or atom. By considering the interactions between neighboring nodes, the models enable the prediction of protein dynamics and the identification of functionally relevant motions.

GNM, the simpler of the two models, assumes that the protein structure behaves like a network of interconnected springs, with each spring representing the interactions between neighboring residues. This model captures the global collective motions of the protein and provides insights into its flexibility and stability. By analyzing the vibrational modes of the ENM, it is possible to identify key regions that are critical for allosteric communication, shedding light on the functional impact of mutations or ligand binding.

ANM takes the ENM a step further by considering the anisotropic nature of protein dynamics, accounting for the directional preferences in the motions of individual residues. ANM incorporates information about the protein structure's Hessian matrix, which describes the second derivatives of the potential energy function with respect to atomic displacements. By considering the anisotropic motions of residues, ANM can provide a more detailed and accurate representation of the protein dynamics.

These models have been successfully applied to a wide range of protein systems, enabling to understand the allosteric communication in diverse biological processes such as enzyme catalysis, signal transduction, and protein-ligand interactions. By elucidating the pathways through which allosteric signals propagate, GNMs and ANMs contribute to the development of rational drug design, where the modulation of allosteric sites can be targeted to control protein function.

Figure 11.4 represents a scheme showing the use of ENMs to identify allosteric regions and dynamic interactions between distant regions of the protein molecules. In Sect. 10.5.1, the reported figures show the allosteric hotspots in human hemoglobin.

The coupled motions of distant regions of the protein structure are specifically appropriate to describe the allosteric nature in multimeric chains, where concerted model for allosteric regulation passes by coupled dynamics of regions also belonging to different chains.

The allosteric nature of multimeric chains and protein-protein interactions is a recent field of investigation, which has found a huge development during the COVID-19 pandemic: the spike protein of the SARS-CoV2 , the virus responsible of the pandemic, is a trimer in charge for the interaction with the human receptor ACE2, activating the infection mechanism. The unbound spike protein is present in two states, open and close. The close to open transition is essential to activate the spike-ACE2 interaction.

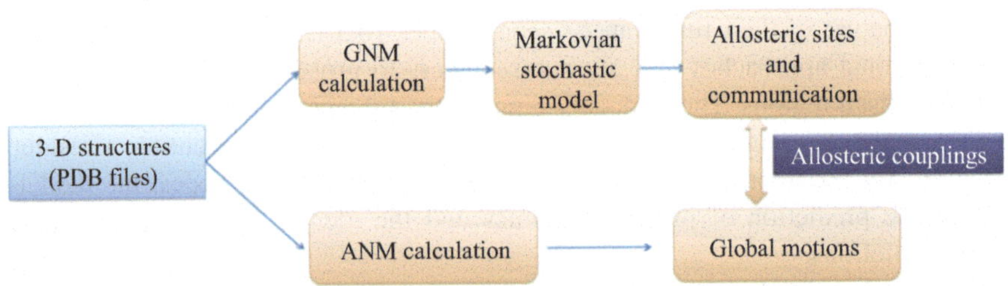

Fig. 11.4 A schematic picture showing the analysis of the allosteric mechanism in proteins by ENMs

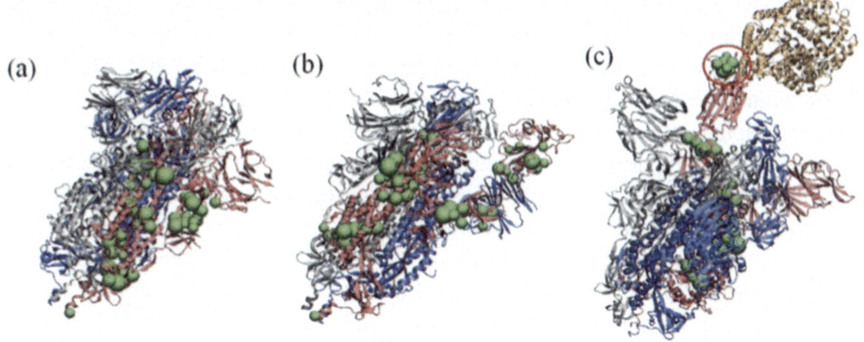

Fig. 11.5 Distribution of the hinge sites (green beads) based on the first three GNM modes: (**a**) closed; (**b**) open; (**c**) bound states. Reprinted with permission from [26]

The active site of the spike-ACE2 interaction is targeted to design purposed drugs. The allosteric interference of the spike-ACE2 interaction is an innovative approach to the therapy development, producing a new class of drugs, referred to as *allosteric drugs* for COVID19 therapy.

In this framework, the ENMs application has provided a precious insight into the allosteric nature of the spike-ACE2 interaction.

Figure 11.5 reports the results of the application of the GNM to identify the Allosteric Modulation Region (AMR) of the spike-ACE2 interaction, which correspond to the hinge sites identified on the basis of the first three GNM modes.

All in all, it is evident that combining Molecular Dynamics of allosteric protein molecules with ENMs can provide a precious perspective on different allosteric mechanisms, with a special focus on those elusive (such as those occurring without a relevant conformation change) or in protein-protein interactions.

11.3.1.3 Energy Transport Analysis of the Allosteric Mechanism

Energy Exchange Networks (EENs) offer a unique perspective on allostery by highlighting the key residues and pathways involved in transmitting information between distal regions. As seen in Sect. 10.5.2, by analyzing the energy flow patterns, it is possible to identify hotspots of allosteric communication and gain insights into how specific residues or mutations affect the transmission of energy and subsequent functional changes. Additionally, these networks can aid in the discovery of allosteric sites and help elucidate the mechanisms by which ligand binding or environmental perturbations propagate through the protein structure.

One of the major advantages of EENs is their ability to quantify the strength of allosteric coupling between different regions of the protein. Through graph-based metrics and computational algorithms, it is possible to give a quantitative description of energy transfer, assess the importance of individual residues or pathways, and predict the impact of mutations or ligand binding on the allosteric behavior. This quantitative analysis provides a deeper understanding of the energetics underlying allostery and facilitates the rational design of targeted interventions for modulating protein function.

Another relevant feature of this computational approach is that EENs can be integrated with experimental techniques such as NMR spectroscopy, X-ray crystallography, and MD simulations to validate and refine the computational models. The combination of experimental data and energy flow analysis offers a comprehensive approach to study allostery, bridging the gap between theoretical predictions and experimental observations.

As in previous computational methods, also EENs can describe the allosteric mechanism in protein-protein interactions. Figure shows the relevant residues as for the allosteric regulation of the spike protein—ACE2 (see for more details the same instance in Sect. 11.3.1.2) determined by EENs analysis.

As an example, Fig. 11.6 shows in the spike-ACE2 complex the residues with the most relevant role in transmitting the energy flows throughout the protein molecular structure.

In summary, energy flow networks have emerged as valuable tools for studying allostery in proteins. By capturing the pathways and strength of energy flow, these networks provide a holistic view of the allosteric communication, enabling the identification of key residues, prediction of allosteric sites, and quantitative assessment of allosteric coupling. Integration with experimental techniques further

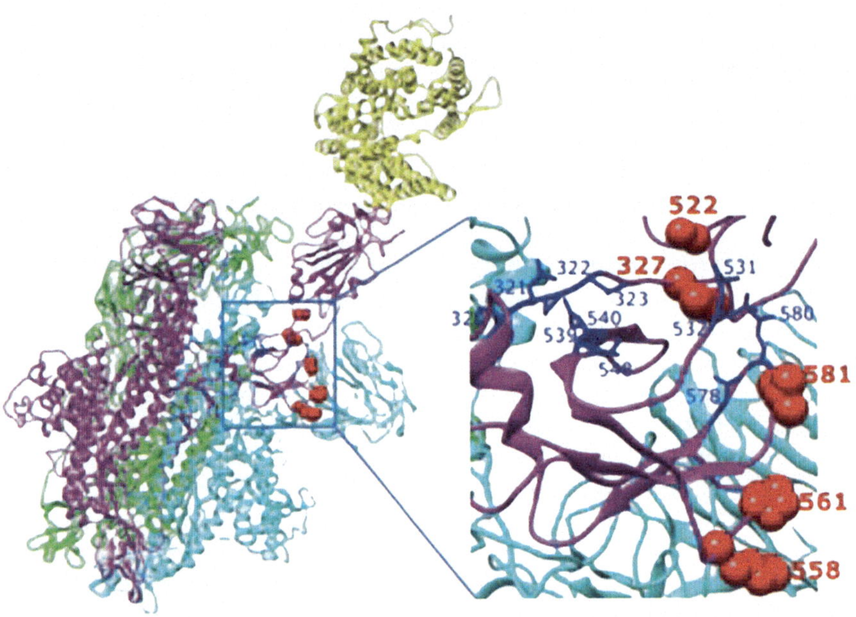

Fig. 11.6 SpikeACE2 complex (chains A, B, and C are in green, cyan, and magenta, respectively, and the ACE2 ectodomain in yellow. The five residues shown as red balls have been identified through EENs application as those with the largest influence on energy transport. Reprinted with permission from [26]

enhances the understanding of allostery, paving the way for future advancements in protein engineering, drug discovery, and the development of novel therapeutic strategies.

11.4 Computational Analysis of Protein-Protein Interactions

Protein-protein interactions (PPIs) are crucial for the functioning of biological systems, as they regulate various cellular processes and are involved in numerous disease pathways. The relevance of PPIs has driven the definition for organisms of *interactome*, i.e., the complete set of physical interactions between molecules in a cell. It includes protein-protein interactions, protein-DNA interactions, protein-lipid interactions, and protein-small molecule interactions. The interactome is a complex network of interactions that is essential for the function of all living cells.

Figure 11.7 shows in a pictorial way the methods to predict the protein-protein interactions.

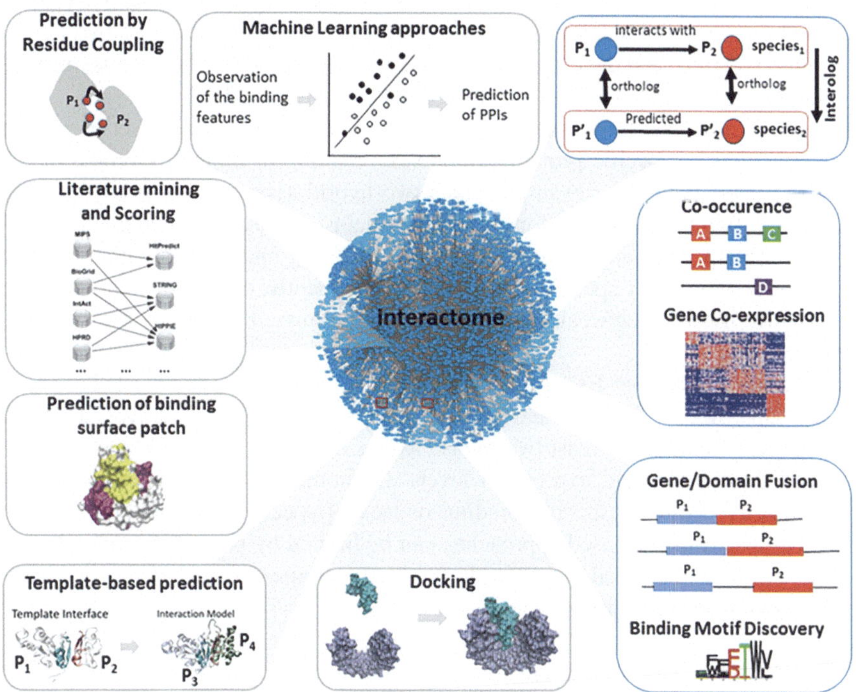

Fig. 11.7 A pictorial perspective of methods to predict the PPIs. Reprinted with permission from [30]

Experimental methods, such as yeast two-hybrid assays, co-immunoprecipitation, and mass spectrometry, have long been employed to study protein-protein interactions (PPIs).

These methods offer direct, *experimental evidence of physical interactions* between proteins and describe the composition and dynamics of protein complexes. However, they show some limitations compared to computational methods:

1. *Time-consuming*: Experimental methods for PPI determination often involve labor-intensive procedures and require considerable time to obtain results. For example, yeast two-hybrid assays involve the construction and transformation of multiple yeast strains, followed by time-consuming growth selection and analysis of protein interactions. Similarly, co-immunoprecipitation experiments require extensive sample preparation, antibody incubations, and subsequent detection methods. These time requirements impair the scalability and efficiency of experimental approaches.
2. *Expensive*: Experimental techniques can be costly, requiring specialized reagents, equipment, and technical expertise. For instance, yeast two-hybrid assays often require specific yeast strains, reporter systems, and reagents for selective growth conditions. Co-immunoprecipitation experiments need specific antibodies, purification columns, and the use of analytical detection methods. These costs limit the number of PPIs that can be investigated within a given budget.
3. *Limited coverage*: Experimental methods typically focus on a subset of protein interactions or specific protein complexes, resulting in limited coverage of the entire interactome. For example, yeast two-hybrid assays are primarily suitable for detecting binary interactions between proteins, while co-immunoprecipitation and mass spectrometry methods excel at identifying interactions within specific complexes or under specific conditions. Consequently, certain interactions, particularly transient or weak interactions, may be missed or overlooked by these techniques.
4. *Technical challenges*: Experimental methods may encounter technical challenges that affect their reliability and accuracy. For example, false positives and false negatives can arise in yeast two-hybrid assays due to potential nonspecific interactions or inadequate expression levels. Co-immunoprecipitation experiments may suffer from nonspecific binding or lack of specificity of antibodies used. Mass spectrometry-based approaches can be limited by sensitivity, the dynamic range of detection, and the difficulty of analyzing membrane proteins or highly heterogeneous complexes.

To overcome these limitations, computational methods have gained prominence in complementing experimental techniques. Computational approaches can provide predictions and insights into potential PPIs, guiding experimental design, and focusing resources on the most promising interactions. Furthermore, computational methods have the advantage of being rapid, cost-effective, and capable of analyzing large-scale interaction networks, thereby enhancing our understanding of the complex protein interactome.

In recent years, computational methods have emerged as valuable tools to complement experimental techniques, enabling the prediction and determination of PPIs in a rapid, cost-effective, and systematic manner.

When studying protein-protein interactions, it is indeed important to understand which proteins or peptides bind to a target protein. This information helps in elucidating the molecular mechanisms underlying various biological processes and can provide insights into the functions of proteins within cellular systems.

To identify proteins or peptides that bind to a target protein, there are several experimental and computational techniques available. Experimental techniques include methods such as yeast two-hybrid assays, co-immunoprecipitation, affinity chromatography, and surface plasmon resonance, among others. These techniques allow researchers to detect and characterize protein-protein interactions by directly observing physical associations between proteins.

On the computational side, various bioinformatics and computational biology tools and algorithms can predict protein-protein interactions based on sequence analysis, protein structure prediction, and molecular docking simulations. These methods use the known properties of proteins and their interactions to make predictions about potential binding partners.

Once a protein-protein complex is given, and its structure is available, there are several ways to quantify the protein-protein interactions and provide a quantitative description of the protein-protein interface. Here are a few commonly used approaches:

1. *Binding Affinity*: The binding affinity measures the strength of interaction between proteins and is typically quantified by the dissociation constant (K_d) or the binding constant (K_a). Experimental techniques such as isothermal titration calorimetry (ITC) or surface plasmon resonance (SPR) can directly measure the binding affinity.
2. *Interface Analysis*: The protein-protein interface can be analyzed at the atomic level to identify key interacting residues and characterize the nature of interactions (e.g., hydrogen bonds, hydrophobic contacts). Various computational tools like PISA, NACCESS, or PDBePISA can calculate parameters like buried surface area, solvent-accessible surface area, and shape complementarity to quantify the protein-protein interface.
3. *Interaction Energy*: MD simulations and energy calculations can provide insights into the energy landscape of the protein-protein complex. Methods, such as molecular mechanics force fields or quantum mechanics calculations, can estimate the interaction energies and contribute to a quantitative understanding of the protein-protein interface.
4. *Binding Free Energy*: Computational methods like molecular dynamics simulations combined with free energy calculations (e.g., molecular mechanics Poisson-Boltzmann surface area, or MM/PBSA) can predict the binding free energy of a protein-protein complex. These calculations provide a thermodynamic view of the interaction and allow for quantitative comparisons between different complexes.

It's important to note that these methods provide approximations and should be interpreted cautiously. Experimental validation and a combination of approaches often provide the most reliable and comprehensive understanding of protein-protein interactions and their quantitative description.

11.4.1 Theoretical Analysis of Protein-Protein Interfaces

The prediction of protein-protein interfaces is based on several theoretical principles, encompassing several features of protein structure, dynamics, and evolutionary conservation. As follows are some key theoretical principles that underpin the prediction of protein-protein interfaces:

- *Structural Complementarity*: Protein-protein interactions often involve complementary shapes and surface properties between the interacting proteins. The principle of structural complementarity suggests that the interacting surfaces of proteins fit together like a lock and key, forming specific contacts and interactions. This principle is based on the observation that protein-protein interfaces tend to have a higher degree of shape complementarity and specific patterns of amino acid interactions compared to noninteracting regions. The structural complementarity is also explored on the basis of domain-domain matching, which leverages in predicting PPIs the modular nature of proteins and the interactions between their constituent domains matching
- *Conservation of Interface Residues*: Functionally important residues at protein-protein interfaces are often evolutionarily conserved. The principle of sequence conservation assumes that residues involved in protein-protein interactions tend to be more conserved across different species or within a protein family compared to noninteracting residues. By identifying conserved residues, sequence-based methods can predict potential interface regions.
- *Energetic Contributions*: The principle of energetics suggests that protein-protein interactions involve favorable energy contributions from multiple types of interactions. These interactions include van der Waals interactions, electrostatic interactions, hydrogen bonding, and hydrophobic interactions. The energetics of protein-protein interfaces can be quantified using computational methods, such as molecular mechanics force fields or knowledge-based potentials, which estimate the favorable and unfavorable contributions to the overall interaction energy.
- *Solvent Accessibility*: Protein-protein interfaces often exhibit distinct solvent accessibility patterns. The principle of solvent accessibility states that residues at the protein-protein interface tend to have intermediate solvent accessibility, as they are partially exposed to the solvent while also being involved in protein-protein contacts. By analyzing the solvent accessibility of residues, computational methods can distinguish interface residues from non-interface residues.

- *Residue Propensity*: Certain amino acid residues have a higher propensity for participating in protein-protein interactions. The principle of residue propensity suggests that specific amino acids, such as arginine, lysine, glutamic acid, and tryptophan, are more frequently found at protein-protein interfaces compared to other residues. This observation has been utilized in developing computational methods that predict interface residues based on residue propensity scores.
- *Binding Hotspots*: Protein-protein interfaces often contain localized regions called binding hotspots that contribute significantly to the overall interaction energy. The principle of binding hotspots suggests that a small subset of interface residues plays a crucial role in the binding affinity and specificity of the interaction. Computational methods aim to identify these hotspot residues by analyzing various features, such as energetics, solvent accessibility, and evolutionary conservation.

These theoretical principles provide a foundation for the development of computational methods and algorithms to predict protein-protein interfaces.

The first essential step to describe PPIs is to identify the type of interaction, on the basis of the classification shown in Fig. 11.8. The first level of PPIs' classification describes the composition of the complexes: the *homo-oligomers* are composed of identical protomers, whereas hetero-oligomers are made up of different protomers. Homo-oligomers can be assembled in an *isologous* or *heterologous* way: in the isologous homo-oligomers, the binding sites in the identical protomers are the same, while in the heterologous assemblies, the binding surfaces are different in the two chains.

Another level of classification is based on the distinction between *obligate* and *non-obligate* interactions. The obligate interactions describe multimers composed of protomers that are not stable on their own but are stable as complexes. There are also obligate complexes from a functional point of view: the activity is granted only by complexes, even if the active site is only on one protomer, otherwise inactive.

The classification of protein-protein interactions can ultimately depend on the length of time they persist. Let's indicate with A and B the two binding proteins; the AB complex forms upon a reversible reaction reaching the thermodynamic equilibrium:

$$A + B \underset{k_{off}}{\overset{k_{on}}{\rightleftharpoons}} AB \tag{11.1}$$

where k_{on} is the kinetic constant of the direct reaction producing the AB complex through the assembly of proteins A and B, whereas k_{off} is the kinetic constant of the inverse reaction, leading to the isolated proteins A and B from the complex disassembly. The dissociation equilibrium constant of Eq. 11.1 is defined as:

$$K_d = \frac{[A] \cdot [B]}{[AB]} = \frac{k_{off}}{k_{on}} \tag{11.2}$$

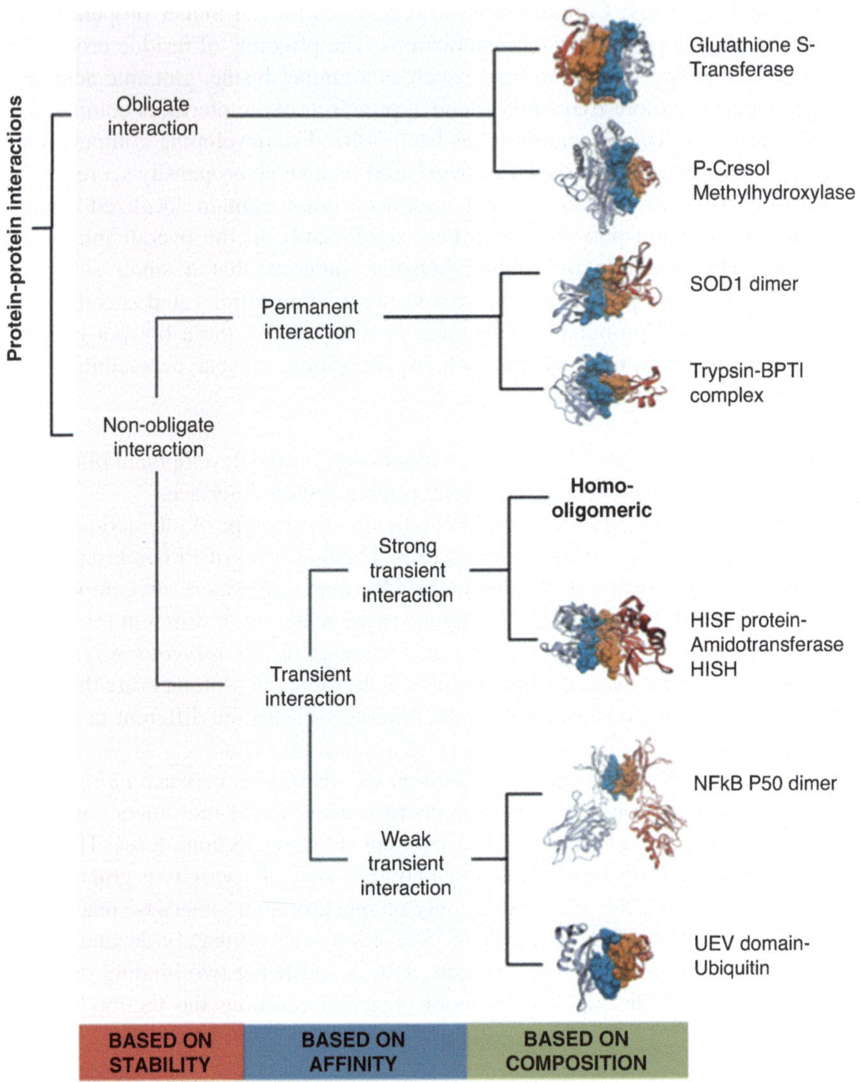

Fig. 11.8 Classification of Protein-Protein Interactions. Reprinted with permission from [30]

being $[A]$, $[B]$ and $[AB]$ the molar concentrations of the proteins, free and bound. Introducing the total concentrations of A and B, $[A]_T = [A] + [AB]$ and $[B]_T = [B] + [AB]$ of the A and B species, respectively, it is possible to derive the fraction of bound A, for instance, as:

$$\theta = \frac{[AB]}{[A]_T} = \frac{[B]}{K_d + [B]} \tag{11.3}$$

If more than one B is bound to A, it is possible to define the protein-protein interaction with the *Scatchard plot*, analogously to protein-ligand binding; indicating with $[B]_b$ the concentration of bound B, the Scatchard plot allows to determine the number n of bound B for each A (*coordination number*) and the equilibrium constant K_d, according to the equation:

$$\frac{[B]_b}{[B]} = -\frac{[B]_b}{K_d} + n \cdot \frac{[A]_T}{K_d} \tag{11.4}$$

The binding affinity finds its thermodynamic definition in the *Gibbs free energy of dissociation* ΔG_d, linked to the dissociation equilibrium constant K_d through the van't Hoff equation:

$$\Delta G_d = -RT \cdot \log \frac{K_d}{c_0} = \Delta H_d - T\Delta S_d \tag{11.5}$$

where c_0 is the standard state concentration (by convention set to $c_0 = 1\ \frac{mol}{L}$), while ΔH_d and ΔS_d are the enthalpy and the entropy of the dissociation, respectively. The *Gibbs free energy of association* $\Delta G_a = -\Delta G_d$ includes two terms, as follows:

$$\Delta G_a = \Delta G_{bond} + \Delta G_{nbond} \tag{11.6}$$

describing two different aspects of the complex formation: the ΔG_{bond} represents the contribution of the bonds between residues in the complex interface, while ΔG_{nbond} includes all "non-bonded contributions" to complex formation, mainly of entropic nature.

An alternative theoretical approach for characterizing protein-protein interactions involves explicitly considering the effect of the solvent environment in complex formation. To begin, we consider an assembly $\mathbf{A} = (A_1, A_2, ..., A_n)$ consisting of n subunits A_i. The standard free energy of dissociation, denoted as $\Delta G^0_{DISS}(A)$, is defined as follows:

$$\Delta G^0_{DISS}(A) = -\Delta G_{int}(A) - T\Delta S^0_{DISS}(A) \tag{11.7}$$

here, $\Delta G_{int}(A)$ represents the binding energy of subunits A_i (with an enthalpic nature), and ΔS^0_{DISS} represents the entropy change upon dissociation.

Negative values of ΔG^0_{DISS} indicate unstable complexes. The binding energy of the subunits $\Delta G_{int}(A)$ in a solvent solution is given by:

$$\Delta G_{int}(A) = \Delta G_{SOLV}(A) - \sum_{i=1}^{n} \Delta G_{SOLV}(A_i) + \sum_{j>i} \Delta G_{CONT}(A_i, A_j) + \sum_{j>i} \Delta G_{ES}(A_i, A_j) \tag{11.8}$$

In this equation, $\Delta G_{SOLV}(A)$ represents the change in free energy of solvation for the entire complex, $\Delta G_{SOLV}(A_i)$ represents the change in free energy of solvation for the individual subunit A_i, and $\Delta G_{CONT}(A_i, A_j)$ and $\Delta G_{ES}(A_i, A_j)$, respectively, account for the contact-dependent interactions (mainly hydrogen bonds) and the electrostatic contributions between the subunits A_i and A_j.

Moreover, the solvation free energy gain upon the formation of an interface between two subunits A_i and A_j, denoted as $\Delta G_{SOLV}(A_i, A_j)$, represents the difference in total solvation energies between isolated and interfacing structures. Negative values of $\Delta G_{SOLV}(A_i, A_j)$ correspond to hydrophobic interfaces or indicate a positive protein affinity.

The solvation energy for a macromolecular assembly A is computed by considering solvation after the formation of a cavity C with a volume V within the solvent environment:

$$\begin{aligned} \Delta G_{SOLV}(A) = \Delta G_{CONT}(A, V) - \Delta G_{CONT}(C, V) \\ + \Delta G_{ES}(A, V) - \Delta G_{ES}(C, V) \end{aligned} \tag{11.9}$$

Here, V represents the bulk solvent, and all energies are computed as differences between the vacuum and solvent environments. The terms $\Delta G_{CONT}(A, V)$ and $\Delta G_{CONT}(C, V)$ encompass the contact-dependent interactions between the bulk solvent V, the macromolecular assembly A, and the cavity C.

The contact-dependent free energy change between two subunits is defined as:

$$\Delta G_{CONT}(A_i, A_j) = N_{hb}^{ij} \cdot E_{hb} + N_{sb}^{ij} \cdot E_{sb} + N_{db}^{ij} \cdot E_{db} \tag{11.10}$$

Here, N_{hb}^{ij}, N_{sb}^{ij}, and N_{db}^{ij} represent the respective numbers of hydrogen bonds, salt bridges, and disulfide bonds between the two subunits A_i and A_j.

The primary contribution to $\Delta G_{DISS}(A)$ is the entropic term $T\Delta S_{DISS}(A)$, which corresponds to the loss of entropy of the solvent upon complex formation due to the decreased mobility of bound solvent molecules.

By explicitly considering solvents and their role in the system, this approach provides insights into the energetics and dynamics of protein complexes. Understanding these factors is crucial for unraveling the mechanisms underlying protein-protein interactions and their significance in biological processes.

At last, in the theoretical framework of protein-protein interactions, a crucial aspect is to understand the contribution of individual residues within the protein-protein interface toward the overall interaction energy. Similar to protein folding and allostery, there exists a significant disparity in the roles played by different residues.

Some residues are primarily responsible for driving the interactions, called *hotspots*, while others make only minor contributions. The latter group may be located in close proximity to the main driving residues due to their proximity to the interface rather than actively participating in the interaction. This disparity in the contributions of individual residues highlights the complexity of protein-protein interactions and emphasizes the need to discern the key residues from those that play a more peripheral role.

The energy contributions of residues in protein complex interfaces exhibit a significant degree of variation, which reflects the presence of cooperativity during the formation of these complexes. *Cooperativity* vplays a crucial role in shaping the energetics of protein interactions. It arises from the collective effect of multiple residues working together to stabilize the complex structure and drive the binding process.

In protein complexes, different residues within the interface can contribute to the binding energy to varying extents. *Hotspots*, i.e., residues with a remarkable role in protein-protein interactions, provide a huge share of the interface energy, while others have a more modest impact. This large variance in energy contributions highlights the interdependence and cooperative nature of the interactions involved in complex formation.

Cooperativity in protein complexes emerges from various factors, including the formation of hydrogen bonds, electrostatic interactions, hydrophobic contacts, and other non-covalent interactions between residues. These interactions can lead to the propagation of structural and energetic changes throughout the complex, influencing the stability and specificity of the protein-protein interaction.

Understanding and quantifying the cooperativity within protein complexes is vital for deciphering the mechanisms underlying their formation and stability. Computational methods, such as MD simulations and free energy calculations, can provide insights into the cooperative behavior of protein interfaces by analyzing the energetics and dynamics of the complex formation process.

11.4.2 Computational Approach to the Prediction and the Analysis of Protein-Protein Interactions

The computational approaches to the description of the protein-protein interactions are based on a general principle that underlies all biological mechanisms: the *biomolecular recognition*. The conceptual description of biomolecular recognition relies on three different models, as shown in Fig. 11.9.

The *lock-and-key model* requires a stiff molecular recognition between the two chains that show an almost complete matching of the interfacing molecular surfaces, both from a spatial and a chemical point of view. This kind of interactions is quite rare and often corresponds to the highest energy interface. In other words, in the lock-and-key model, biomolecular recognition is explained by the complementary shapes of a receptor (lock) and a ligand (key). The complementary shape and interactions between the two chains lead to a stable complex with lower interface energy.

The *induced fit model* describes a protein-protein interactions provoking a conformation changes in interacting protein molecules. Therefore, the induced-fit model in protein-protein interactions does provide insights into the allosteric nature of these interactions. Indeed, these conformation changes can result in altered pro-

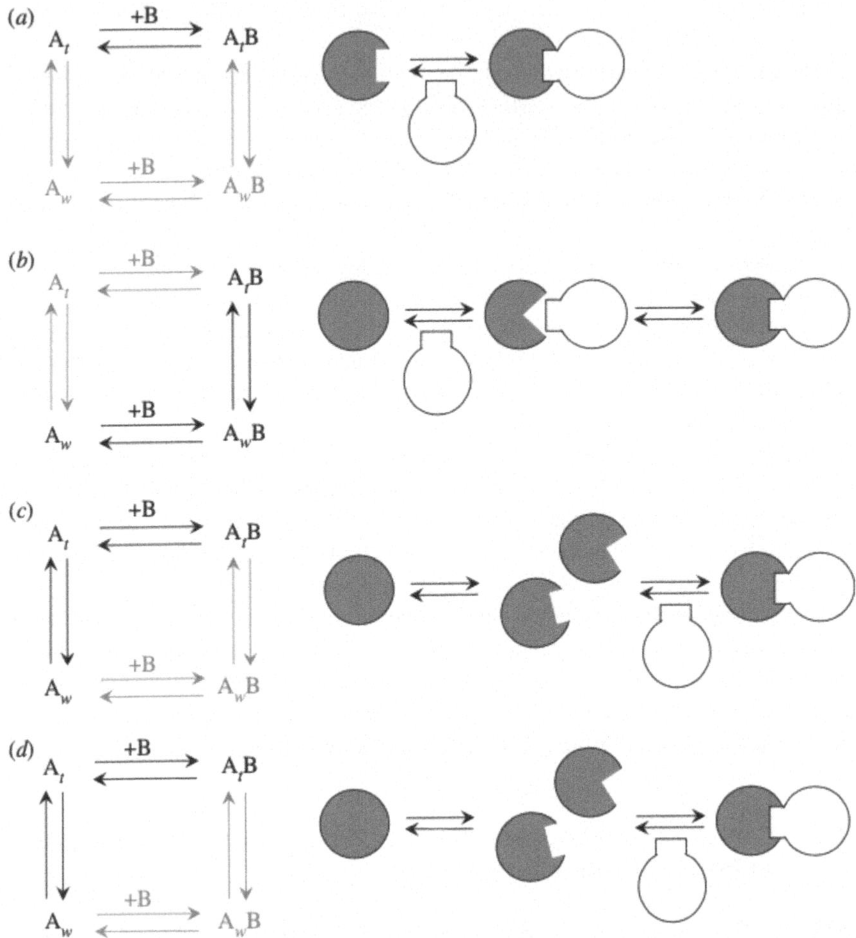

Fig. 11.9 The concept of biomolecular recognition encompasses three models: (**a**) lock-and-key; (**b**) induced-fit; (**c**) conformation selection (also known as dynamical fit). The protein's tight conformation, denoted as A_t, represents the binding competent state, while the weak conformation, denoted as A_w, signifies the binding incompetent state. In this simplified scheme, we assume that protein B does not undergo any conformation changes, whereas in reality, in most cases, it is the opposite. Reprinted with permission from [29]

tein structure and dynamics, leading to functional changes in the protein's activity or binding properties. This dynamic response to binding is a key characteristic of allosteric regulation.

In protein-protein interactions, the induced fit model can describe how the binding of one protein to another induces conformation changes in both proteins, leading to the formation of a stable complex and modulating their biological function. These conformation changes can occur through alterations in side-chain packing, loop movements, or domain reorientation, among other mechanisms.

Finally, a third mechanism is described by the *dynamic fit model*, also known as *conformation selection*. This model requires all interacting species are preexisting in multiple configurations (*configurational ensemble*), but only one configuration embodies the best fitting, i.e., resulting in the most stable context. So the energy of the complex in a way select the most appropriate conformation for the interaction, shifting the equilibrium toward the complex formation.

The conformation selection model enriches the insights on the allosteric nature of PPIs. In the context of allosteric interactions in PPIs, the conformation selection model suggests that the binding of one protein to another can selectively stabilize a conformation of the target protein that is competent for binding and signal transduction. This can lead to the transmission of information across the protein structure, affecting the activity or binding of other regions.

In the conformation selection model, the binding event does not induce large conformation changes but rather selects and stabilizes an existing conformation state. This is in contrast to the induced-fit model, where binding induces conformation changes. Both models provide different perspectives on how allosteric interactions occur in PPIs, with the conformation selection model focusing on the equilibrium of preexisting conformations and their selective stabilization upon binding.

Figure 11.10 sketches the computational approach to analyze protein-protein interactions. If the interface is not known, it is necessary to implement and apply methods to identify the putative interface. The analysis of the interface implies the identification of key parameters, which is the shape and extent of the interface, the involved residues and their ranking, to sort out the ones mostly responsible of the whole interactions (*hotspots*).

If the protein-protein interface is unknown, its identification requires two main approaches:

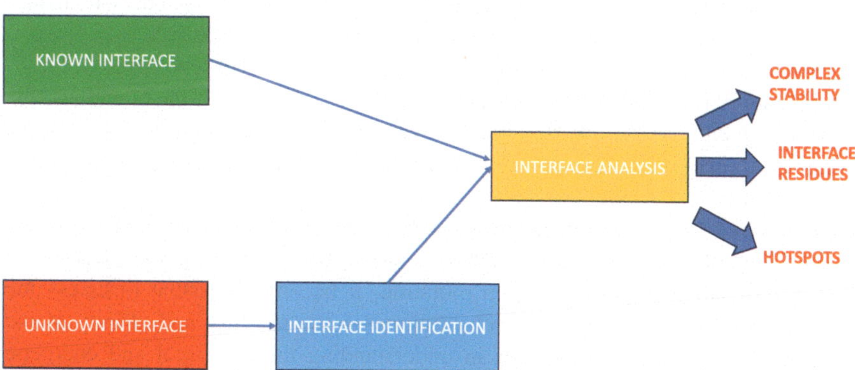

Fig. 11.10 The computational approach to the protein-protein interaction analysis

- *gene/sequence approach*: the first attempt is to predict protein-protein interface on the base of similar protein sequences with known complex structures; computational methods include phylogenetic trees (ibs2analyzer), sequence-based method (PrePPI) and gene/protein association (STRING);
- *structure-based approach*: the structure-based approaches to the identification of protein-protein interfaces rely on analyzing the three-dimensional structures of proteins to identify key elements of the interaction. Once the structural information of the two interacting protein is known, molecular docking and molecular dynamics simulations to attempt to predict the binding mode of two or more proteins based on their individual structures and known constraints, providing a full description of the functional and dynamic features of the complex. In addition to the multipurpose HADDOCK web server, mentioned in Sect. 8.3, ClusPro, a widely used web server for protein-protein docking, employs a two-step approach that involves initial rigid-body docking followed by the refinement of the complex using an energy-based algorithm. ClusPro can handle a wide range of protein-protein complexes and provides users with ranked models of the predicted protein-protein interactions.

Molecular thermodynamics plays a crucial role in the quantitative description of protein-protein interfaces. It provides a framework for understanding the thermodynamic properties and behavior of molecules at the molecular level, in different ways, as follows:

- *Binding Free Energy*: Molecular thermodynamics allows for the calculation of binding free energy, which quantifies the stability and strength of protein-protein interactions. The estimation of the binding free energy changes associated with the formation of the protein-protein complex is carried out including the energetic contributions from various molecular interactions, such as electrostatics, van der Waals forces, and solvation effects
- *Thermodynamic Parameters*: Molecular thermodynamics enables the calculation and interpretation of various thermodynamic parameters associated with protein-protein interfaces. These parameters include enthalpy (H), entropy (S), Gibbs free energy (G), and the equilibrium constant (K_{ass}) between the bound and the unbound states. These thermodynamic parameters provide quantitative insights into the energetic contributions and driving forces behind the protein-protein interaction.
- *Mutational Analysis*: Molecular thermodynamics quantifies the effects of amino acid mutations on protein-protein interfaces. By calculating the changes in thermodynamic parameters upon mutation, it becomes possible to assess the impact of specific amino acid substitutions on the stability and affinity of the complex. This information aids in understanding the structural and energetic determinants of the protein-protein interaction.
- *Experimental Design and Optimization*: Molecular thermodynamics provides a quantitative framework for experimental design and optimization of protein-protein interactions. It allows for the prediction and evaluation of the effects of

experimental conditions, such as temperature, pH, and salt concentration, on the thermodynamic behavior of the system. This knowledge can guide experimental setups and assist in optimizing conditions for studying and manipulating protein-protein interfaces.

Molecular thermodynamics contributes to the quantitative description of protein-protein interfaces by enabling the calculation of binding free energies, equilibrium constants, and other thermodynamic parameters. It facilitates the analysis of mutational effects, guides experimental design, and provides insights into the energetics and stability of protein-protein interactions.

There are several web servers focused on the quantitative analysis of protein-protein interfaces, such as:

1. PDBePISA (Protein Data Bank in Europe—Protein Interfaces, Surfaces, and Assemblies): PDBePISA provides a web-based interface for the analysis of protein-protein interfaces, surfaces, and assemblies. It calculates various parameters related to the protein-protein interface, including buried surface area, hydrogen bonds, and detailed interaction networks.
2. PRODIGY (Protein-Protein Binding Affinity Prediction): PRODIGY is a web server that predicts the binding affinity of protein-protein complexes. It is developed within the HADDOCK (see Sect. 8.3). It employs an empirical scoring function based on a combination of structural and evolutionary features to estimate the dissociation constant (K_{diss}) of the complex. PRODIGY provides quantitative information about the binding affinity and can be useful in understanding the strength of protein-protein interactions.
3. ITScore-PP: ITScore-PP is a web server for the quantitative analysis of protein-protein interfaces based on the concept of interaction density. It calculates the interface score, which represents the propensity of a protein-protein interface to interact. The interface score is derived from the spatial distribution of atoms and residues and can be used to compare and analyze different protein-protein interfaces.
4. COCOMAPS (COmplex COntact MAPS): COCOMAPS is a web server that analyzes protein-protein interfaces based on contact maps. It provides quantitative information about interfacial residues, contact density, and pairwise interaction energies. COCOMAPS allows for the visualization and exploration of protein-protein interfaces and their structural features.

These web applications have its own set of features, algorithms, and outputs, so it is beneficial to explore and compare the capabilities of multiple servers to find the most suitable one for specific research needs.

Finally, a very crucial topic in the PPI study is to identify *hotspots*, i.e., residues which play a major role in the protein-protein binding energy. As seen above, PPIs hotspots are specific regions within the protein-protein interface that contribute significantly to the binding affinity and stability of the complex. These hotspots often involve clusters of residues that make crucial contacts with the partner protein or contribute a large portion of the binding energy. Identifying PPI hotspots is essential for understanding the key determinants of protein-protein interactions and for designing targeted interventions.

As follows, a list a few commonly used computational approaches:

- *Alanine Scanning Mutagenesis*: In alanine scanning mutagenesis, each residue at the protein-protein interface is systematically mutated to alanine, and the impact of these mutations on the binding affinity is assessed. Residues that, when mutated to alanine, significantly decrease the binding affinity are considered hotspot residues. Computational alanine scanning methods use MD simulations or free energy calculations to estimate the changes in binding energy upon alanine mutation of each residue at the interface. These methods can efficiently identify hotspot residues without the need for extensive experimental mutagenesis.
- *Energy-Based Methods*: Energy-based methods utilize various scoring functions or energy potentials to evaluate the contribution of individual residues to the binding energy. These methods typically involve molecular dynamics simulations or molecular mechanics calculations to estimate the energy of the protein-protein complex. Residues with higher contributions to the binding energy are identified as hotspot residues.
- *Machine Learning Approaches*: Machine learning methods have been developed to predict hotspot residues based on sequence and/or structural features. These methods utilize training datasets that contain experimental or computationally derived information about hotspot residues and use this information to build predictive models. Machine learning algorithms, such as random forests, support vector machines (SVM), or neural networks, are commonly employed in this context.
- *Solvent Mapping*: Solvent mapping methods analyze the solvent-accessible surface of the protein-protein complex to identify regions that are highly occupied by water molecules. Residues in these regions are more likely to contribute to the binding energy and are considered potential hotspots.

It's important to note that a combination of these computational methods, along with experimental validation, can provide more accurate and reliable hotspot predictions. Additionally, the availability of web servers and software tools that implement these methods allows researchers to easily perform hotspot analysis for their protein-protein complexes.

As follows are few examples of web servers providing computational identification of hotspots:

1. KFC2: Knowledge-based FADE and Contact Analysis (KFC2); KFC2 is a web server for hotspot prediction in protein-protein interfaces. It utilizes a knowledge-based statistical potential to identify energetically important residues at the interface. KFC2 accepts protein structures in PDB format and provides hotspot predictions along with interactive visualizations.

2. PIIMS: PIIMS is a web server providing information about hotspots by calculating the binding free energy changes upon alanine mutation scanning. In addition, the mutation scanning helps understanding the impacts of hotspot mutations:
3. SMMPPI: SMMPPI is a web server approaching a comprehensive machine-learning analysis of PPIs, including the identification of hotspots. The main focus of the tool is to identify modulators of PPIs, and it provides a plenty of information on the energy change upon modulators' binding.
4. ECMIS: ECMIS leverages features of protein structure and sequence space at residue level to identify PPI's hotspot residues. The method is quite accurate to identify also residues contributing on a small scale, providing a clear and comprehensive perspective on the PPI's interface.

Finally, an additional important support for the hotspots analysis and identification is provided by specialized databases, which collect hotspots and, more in general, PPI's information coming from experiments. The sharing of this information is crucial for different reasons, for training machine-learning algorithms and, more in general, for comparison and accuracy assessment of prediction methods. Here as follows same databases freely accessible on the web:

1. *PPI-Hotspot*^{DB}: PPI-Hotspot^{DB} collects data on complex structures and hotspots, determined experimentally; the database has been applied to create a nonredundant benchmark, PPI-Hotspot^{DB} + PDB, providing a tool to assess methods for predicting PPI-hotspots using the free structure as input.
2. BioGRID: BioGRID (Biological General Repository for Interaction Datasets): BioGRID is a curated database that collects and stores PPIs data from the scientific literature. It includes both physical interactions and genetic interactions, covering a wide range of organisms.

These databases and resources offer valuable information and tools for identifying, predicting, and analyzing hotspots in protein-protein interfaces.

Additional Resources and Recommended Literature

A general textbook for protein-ligand binding is [12]. A general perspective on protein-protein interactions can be found in [15]. A general framework on protein interactions is provided in [8].

Finally, an updated general perspective on computational methods for allostery analysis is provided in [7].

Bibliography

Books

1. Alavi S (2020) Molecular simulations: fundamentals and practice. Wiley, New York
2. Bailey JE, Ollis DF et al (2018) Biochemical engineering fundamentals. McGraw-Hill, New York
3. Buxbaum E et al (2007) Fundamentals of protein structure and function, vol 31. Springer, Berlin
4. Cantor CR, Schimmel PR (1980a) Biophysical chemistry: Part III: the behavior of biological macromolecules. Macmillan, New York
5. Cantor CR, Schimmel PR (1980b) Biophysical chemistry: Part III: the behavior of biological macromolecules. Macmillan, New York
6. Chandler D (1987) Introduction to modern statistical mechanics. Oxford University Press, Oxford
7. Di Paola L, Giuliani A (2021) Allostery: methods and protocols. Springer US, Berlin
8. Gromiha MM (2020) Protein interactions: computational methods, analysis and applications. World Scientific, Singapore
9. Hill TL (1986) An introduction to statistical thermodynamics. Courier Corporation, North Chelmsford
10. Hiroaki K (2001) Foundations of systems biology. The MIT Press, New York
11. Israelachvili, JN (2011) Intermolecular and surface forces. Academic press, Cambridge
12. Mannhold R, Kubinyi H, Folkers G (2012) Protein-ligand interactions. Wiley, New York
13. Misra G (2017) Introduction to biomolecular structure and biophysics: basics of bio-physics. Springer, Berlin
14. Nelson DL, Lehninger AL, Cox MM (2008) Lehninger principles of biochemistry. Macmillan, New York
15. Poluri KM, Gulati K, Sarkar S (2021) Protein-protein interactions: principles and techniques, vol I. Springer Nature, Berlin
16. Prausnitz JM, Lichtenthaler RN, De Azevedo EG (1998) Molecular thermodynamics of fluid-phase equilibria. Pearson Education, London
17. Rapaport DC (2004) The art of molecular dynamics simulation. Cambridge University Press, Cambridge
18. Sandler SI (2017) Chemical, biochemical, and engineering thermodynamics. Wiley, New York
19. Tramontano A (2006) Protein structure prediction: concepts and applications. Wiley, New York
20. Voit E (2017) A first course in systems biology. Garland Science, New York

L. Di Paola, *Fundamentals of Molecular Bioengineering*,
https://doi.org/10.1007/978-3-031-42022-1

Articles

21. Böde C et al (2007) FEBS Lett 581(15):2776–2782
22. Cumbo F et al (2014) PLoS One 9(10):e105001
23. Di Paola L, Giuliani A (2015) Curr Opin Struct Biol 31:43–48
24. Di Paola L et al (2013) Chem Rev 113(3):1598–1613
25. Di Paola L et al (2022) Entropy 24(7):998
26. Hadi-Alijanvand H et al (2022) ACS Omega 7(20):17024–17042
27. Hu G (2021) Allostery: Methods Protoc 2253:21–35
28. Karplus M, Kuriyan J (2005) Proc Natl Acad Sci 102(19):6679–6685
29. Kastritis PL, Bonvin AMJJ (2013) J R Soc Interface 10(79):20120835
30. Keskin O, Tuncbag N, Gursoy A (2016) Chem Rev 116(8):4884–4909
31. Kuhlman B, Bradley P (2019) Nat Rev Mol Cell Biol 20(11):681–697
32. Leitner DM (2008) Annu Rev Phys Chem 59:233–259
33. Liu J, Nussinov R (2016) PLoS Comput Biol 12(6):e1004966
34. Petrey D, Honig B (2005) Mol Cell 20(6):811–819
35. Raskatov JA, Teplow DB (2017) Sci Rep 7(1):12433
36. Segura-Cabrera A et al (2012) Med Chem Drug Des 27:27–54
37. Srivastava A et al (2018) Int J Mol Sci 19(11):3401
38. Strodel B (2021) J Mol Biol 433(20):167182
39. Swain JF, Gierasch LM (2006) Curr Opin Struct Biol 16(1):102–108

Index

Symbols
α-helices, 51
β-sheets, 51
β_2 Adrenergic Receptor, 221
ab initio modeling, 155
3DLigandSite, 234

A
ab initio modeling, 172, 177, 178, 237
ACE2, 186
Acid dissociation equilibrium, 92
Acid dissociation equilibrium constant, 92
Activity, 16, 85
Activity coefficient, 16, 85
Adair model, 121
Adaptability, 188
Adjacency matrix, 202
Adsorption on resins, 108
Alanine Scanning Mutagenesis, 262
Albumin-hepatic toxins, 108
Allosteric Modulation Region, 247
Allostery, 119, 120, 200, 206, 223, 236, 248
AlphaFold, 172
AMBER, 154, 160
Amino acid residue, 49
Amino acids, 48
Amyloids, 170
Andersen Thermostat, 149
Anfinsen's dogma, 166, 176
Anfinsen's experiment, 165
Anfinsen's theory, 165
Anisotropic Network Model, 215, 245
Anisotropic Network Modeling, 213
Average shortest path, 205

B
Berendsen Barostat, 151
Berendsen Thermostat, 149
Betweenness centrality, 192, 205, 240
Binding Free Energy, 260
Binding Site Conservation Theory, 228
Bio3D, 162
Biocompatibility of materials, 108
BioGRID, 263
Biomolecular recognition, 257
BioSimSpace WebApp, 161
Bipartite networks, 194
Boltzmann constant, 23
Boltzmann's law, 24

C
Canonical Ensemble, 34, 147, 151
Canonical partition function, 37, 81
Carnahan-Starling equation, 102
CASP, 173
Centrality, 191
Chaostropic salts, 101
Chaperonins, 59, 108, 167
CHARMM, 154, 161
Chemical graphs, 200
Chemical Potential, 12
Closeness centrality, 192
ClusPro, 260
Clustal Omega, 175
ClustalW, 175
COACH, 233
COACH-D, 236
COCOMAPS, 261
Cohn-Edsall equation, 100
Competitive binding, 118, 119
Complex systems theory, 168, 187, 188, 196, 198
Compressibility, 73
Compressibility factor, 87, 102
ConCavity, 236
Condensation reaction, 49
Configurational ensemble, 259

L. Di Paola, *Fundamentals of Molecular Bioengineering*,
https://doi.org/10.1007/978-3-031-42022-1

Configurational entropy, 45
Configurational free energy, 58
Configurational integral, 81
Configurational properties, 67
Conformation landscape of a protein, 166, 169
Conformation selection, 259
Convolutional Neural Networks, 232
Cooperativity, 64, 121, 219, 257
Cosmotropic salts, 101
Coulomb's law, 69
Coupled motions, 237
Covolume, 89
CURP, 222

D
de Broglie wavelength, 102
Debye-Hückel length, 100
Deep learning, 175, 232, 234
Degree centrality, 192
Degree matrix, 209
Denaturing agents, 59, 62
Density functional theory, 155
Directed graphs, 190
Distance matrix, 202
Disulfide bond, 52
Drug design, 159, 186, 236
Drug discovery, 225, 236
Drug-target interactions, 234
Dynamic fit model, 259
Dynamic Flexibility Theory, 228
Dynamical behavior of complex systems, 194
DynaMine, 183

E
ECMIS, 263
Eigenvector centrality, 193, 206
Elastic Network Model, 213, 217, 239, 245
Electrostatic potential of proteins, 97
Emergence, 188, 196, 197
Empirical potentials, 154
Energetic Hotspots Theory, 228
Energy Exchange Networks, 247
Energy exchange networks, 239
Energy Flow Networks, 213, 217
Energy transport, 217, 237
Ensemble, 31
Ensemble choice in Molecular Dynamics simulations, 146
Enthalpy, 11
Enzyme inhibition, 54
Enzymes, 54
Equation of state, 21, 76
Equilibrium dialysis, 105
Eterotropic allosteric modulator, 120
Euler's theorem, 12
Evolutionary Conservation Analysis, 230
Excess energy, 221

F
Feedback inhibition, 120
Feedback loops, 189, 196, 197
Fibrous proteins, 53
Fiedler vector, 209
FINDSITE, 235
First Postulate of Statistical Thermodynamics, 33
First Principle of Thermodynamics, 5
Fluorescence Resonance Energy Transfer, 218
Fold recognition, 177
Folding Gibbs free energy, 57
FoldX, 184
Force fields, 154
Fragment assembly methods, 172
Fragment-based Approaches, 231
Fugacity, 15
Funnel landscape theory, 167

G
Gō and Gō-like models, 155
Gaussian Network Model, 213, 245, 247
Generalized Ensemble, 152
Generative models, 233
Geometric networks, 192
Gibbs ensemble theory, 31
Gibbs free energy, 11
Gibbs free energy conformation energy, 165
Gibbs free energy of dissociation, 255
Gibbs free energy of solvation, 65, 256
Gibbs-Duhem equation, 13
Globular proteins, 53
Grand Canonical Ensemble, 40
Grand canonical partition function, 40
Graph, 189
Graph Theoretical Approaches, 239
GROMACS, 160
Guimerà-Amaral cartography, 210

H
HADDOCK, 161, 185
Hamiltonian, 142
Hand-in-glove model, 130
Hard-sphere intermolecular potential, 70
Helmotz free energy, 11

Hemoglobin, 126, 215, 216, 241
Hemoglobin-respiratory gases, 108
Henderson-Hasselbach equation, 92
Hessian matrix, 215
Heuristic methods, 172
Hidden Markov models, 176
Hierarchical clustering algorithm, 208
Hierarchical networks, 194
Hierarchy, 189
Hill plot, 114
History and path dependence, 189
HMMER, 176
Hofmeister series, 101
Homology modeling, 172, 176, 199, 230
Homotropic allosteric modulator, 120
HoTS, 234
Hotspot, 228, 243, 244, 253, 256, 259, 261
HybridPBRpred, 234
Hydrophobic effect, 58
Hydrophobicity, 58

I
I-TASSER, 172, 173, 179
ibs2analyzer, 260
Ideality, 70
In-degree centrality, 192
Induced fit model, 55, 130, 228, 257
Induced folding, 169, 170
Information Theoretical Approaches, 239
Interactome, 249
Interconnectivity, 189
Interdependence, 188
Intermolecular forces, 67
Intermolecular potential, 67, 68
Internal Energy, 5
Intrinsically Disordered Proteins, 159
Ionic strength, 98
Isoactivity condition of phase equilibrium, 17
Isoelectric point, 95
Isofugacity condition of phase equilibrium, 17
Isothermal titration calorimetry, 186
Iterative refinement algorithms, 175
ITScore-PP, 261

K
k-means clustering algorithm, 208
Karplus-Levitt model, 157
KFC2, 262
Kinetic energy of a molecule, 22
Kinetic theory of gases, 20
Kinetic traps, 165
Knowledge-based methods, 172

L
Langevin Dynamics, 149
Langmuir isotherm, 113
Laplacian matrix, 209
Law of perfect gases, 70
Leapfrog integration, 144
Lennard-Jones potential, 71, 153
Levinthal paradox, 167
Ligand, 107, 109
Ligand clustering, 229
Ligand Clustering and Scaffold Analysis, 231
Ligand Similarity and Molecular Fingerprints, 231
Ligand-Based Prediction Approaches, 229
Local thermal conductivity, 220
Lock-and-key model, 55, 130, 228, 257

M
Machine learning, 171, 172, 229, 237, 262
MAFFT, 176
Markov State Models, 239
Martyna-Tobias-Klein Barostat, 152
Martyna-Tuckerman-Tobias-Klein (MTTK) Thermostat, 149
Master Equation, 219
McMillan-Mayer theory, 79
MDpocket, 233
MDsrv, 161
MDWeb, 161
Mean-field approximation, 69
MedusaDock, 161
Method of the Maximum Term Approximation, 35
Michaelis-Menten constant, 133
Michaelis-Menten equation, 133
Microcanonical Ensemble, 33, 41, 147, 151
Mie's intermolecular potential, 71, 72
Minimum frustration principle, 167
Misfolded proteins, 170
MODELLER, 179
Modeller, 173
Modular networks, 194
Modularity, 189, 207, 237, 240
Molecular complementarity, 228
Molecular Docking, 199, 229
Molecular Dynamics, 172, 182, 233, 237, 238, 247
Molecular fingerprints, 235
Monod-Wyman-Changeux (MWC) model, 122
Monte Carlo Barostat, 152
Moonlighting proteins, 170
Multiple sequence alignment, 175
Multiple sites ligand protein binding, 119

MUSCLE, 175
Mutational Analysis, 260

N
NAMD, 160
Native protein structure, 51, 57
Natively unfolded proteins, 168
Needleman-Wunsch algorithm, 174, 175
Negative cooperativity, 121
Network, 189, 237
Network clustering, 208, 241, 243, 244
Network descriptors, 191
Network invariants, 191
Network topology, 196
Network-based allostery, 238
Newton's law of motion, 148
Node degree, 192, 203
Non-heuristic methods for folding, 172
Nonlinearity, 188, 196, 197
Normal Mode Analysis, 212, 239
Nosé-Hoover chain, 149
NPT Ensemble, 146, 147, 151, 152
Nuclear Magnetic Resonance (NMR) spectroscopy, 218
NVT Ensemble, 146, 147

O
OpenMM, 161
OPLS, 154
Osmosis, 77
Osmotic coefficient, 85
Osmotic pressure, 78
Osmotic virial coefficients, 87
Out-degree centrality, 192

P
Parrinello-Rahman Barostat, 152
Partial molar volume, 12
Participation coefficient, 210, 244
Pattern and Motif Analysis, 230
PDBePISA, 185, 261
Peptide bond, 49
Perron-Frobenius theorem, 193
Pharmacophore modeling, 229, 231, 235
Phase equilibria, 16
PIIMS, 263
PISA web server, 185
pKa, 92
Plane of Clapeyron, 28
Positive cooperativity, 121
PPI-Hotspot^{DB}, 263
Preferential exclusion of salting-out salt, 105
PrePPI, 260
Primary Structure of proteins, 51
Prions, 108, 171
PRODIGY, 261
ProDy, 216
PROSS 2, 183
Protein contact networks, 201, 239, 240
Protein Data Bank (PDB), 171
Protein denaturation, 52, 54, 57
Protein Engineering and Design, 226
Protein folding, 57
Protein networks, 198
Protein solvation, 65, 181
Protein stability, 180
Protein structure, 49
Protein Structure Network, 217
Protein topology, 191
Protein's folding kinetics, 167
Protein-ligand binding, 107, 223, 227, 236
Protein-protein interaction networks, 193, 198, 199
Protein-protein interactions, 223, 249, 258, 259
PSI-BLAST, 176

Q
Quantitative Structure-Activity Relationship, 231, 235
Quantum mechanics/molecular mechanics (QM/MM) methods, 155
Quantum state, 32
Quaternary structure of proteins, 53

R
Radial distribution function, 73
Random coil, 57
Random networks, 193
Random-Phase Approximation (RPA), 101
RaptorX, 179
Receptor-ligand, 108
Recurrent Neural Networks, 232
Reinforcement learning, 233
Replica Exchange Molecular Dynamics (REMD), 150
Robustness, 188
Rosetta, 172
RosettaDock, 185

S
Salting-in of proteins, 98
Salting-out constant, 100

Salting-out of proteins, 98
SARS-CoV2, 173, 186, 246
Scale-free networks, 194
Scatchard equation, 124
Scatchard plot, 113, 119, 255
Schrödinger equation, 32
SDM, 184
Second Postulate of Statistical Thermodynamics, 33
Second Principle of Thermodynamics, 7
Secondary Structure of proteins, 51
Self-organization, 188, 198
Semiempirical potentials, 154
Sequence alignment, 174, 178
Serum protein-drugs, 108
Shape Complementarity Theory, 228
Shortest path, 192
Shortest path matrix, 192, 240
Signal transmission in networks, 192, 242
Single-Molecule Spectroscopy, 218
SiteMap, 234
Small-world networks, 194
Smith-Waterman algorithm, 174, 175
Soft-sphere intermolecular potentials, 71
Solvation, 70
Solvation layer, 66
Solvation Molecular Dynamics simulations, 146
Solvent Mapping, 262
Spectral clustering, 209, 243
Spike protein, 186
Spike-ACE2 interaction, 246, 248
Square-well intermolecular potential, 72, 88
Standard Gibbs free energy of ionization, 93
Standard Gibbs free energy of reaction, 30
STRING, 260
Structural Biology, 226
Structural Protein Networks, 199
Structure-based Virtual Screening, 229
Substrate, 54
SuperPred, 235
Supersystem, 34
Surface plasmon resonance, 186
Sutherland potential, 71
SwarmDock, 185
SWISS-MODEL, 179
Symplectic space, 148
Systems biology, 187

T
T-Coffee, 175
Tanford's hydrophobicity scale, 58
Tertiary Structure of proteins, 52
Theory of molecular recognition, 227
Thermal Entropy, 45
Thermodynamic equilibrium state, 32
Thermodynamic Probability, 45
Thermodynamic systems, 3
Thermodynamics of unfolding, 57
Time-Resolved Fluorescence Spectroscopy, 218
Transition State Analysis, 151

U
Undirected graph, 190
Unweighted graphs, 190, 202

V
van der Waals equation of state, 88
van't Hoff equation, 31, 60, 116
van't Hoff law for osmotic systems, 70
Verlet integration, 144, 148
Vibrational Analysis, 151
Virial coefficients, 79
Virial equation of state, 79
Virial theorem, 80
VMD, 161

W
Web PSN, 217
Weighted graphs, 191, 202

Z
z-score of the intracluster connectivity, 210
Zwitterion, 94

The manufacturer's authorised representative in the EU is Springer Nature Customer Service Centre GmbH, Europaplatz 3, 69115 Heidelberg, Germany. If you have any concerns regarding our products, please contact ProductSafety@springernature.com

Printed and bound by CPI Group (UK) Ltd, Croydon, CR0 4YY
07/07/2026
02160911-0002